Transition Metal-based Nanofoams for Electrochemical Systems

Manufacturing, Characterization and Applications

Transition Metal-based Nanofoams for Electrochemical Systems

Manufacturing, Characterization and Applications

By

Maria de Fátima Montemor
Universidade de Lisboa, Portugal
Email: mfmontemor@tecnico.ulisboa.pt

Diana M. Fernandes
LAQV/REQUIMTE, University of Porto, Portugal
Email: diana.fernandes@fc.up.pt

Alberto Adan-Mas
University of Lisbon, Portugal
Email: albertoadanmas@gmail.com

Ana Catarina Alves
Universidade de Lisboa, Portugal
Email: caatarina.alves@gmail.com

Gabriel Garcia Carvalho
Universidade de Lisboa, Portugal
Email: gabriel.carvalho@tecnico.ulisboa.pt

and

Inês S. Marques
LAQV/REQUIMTE, University of Porto, Portugal
Email: up201608306@edu.fc.up.pt

ROYAL SOCIETY OF CHEMISTRY

Print ISBN: 978-1-83767-508-1
PDF ISBN: 978-1-83767-511-1
EPUB ISBN: 978-1-83767-512-8

A catalogue record for this book is available from the British Library

The Royal Society of Chemistry is a charity, registered in England and Wales, Number 207890, and a company incorporated in England by Royal Charter (Registered No. RC000524), registered office: Burlington House, Piccadilly, London W1J 0BA, UK, Telephone: +44 (0) 20 7437 8656.

Visit our website at www.rsc.org

Printed in the United Kingdom by CPI Group (UK) Ltd, Croydon, CR0 4YY, UK

Preface

Metal nanoforms are an extraordinary class of inorganic materials with a variety of different chemical compositions that combine the intrinsic physical and mechanical properties of metals with a highly porous nanoarchitecture. As a result, they have a seemingly never-ending set of possible applications. This book covers different aspects regarding metal foams, namely, the synthetic and functionalization methodologies, types of metal foams, the most important characterization techniques and their application in electrochemical systems.

In ***Chapter 1***, the most promising energy and conversion devices are introduced along with the current state-of-the-art materials. Metal foams are also introduced as emerging candidates for these energy devices. In ***Chapter 2***, several preparation routes for metal foams are presented; in particular, templating synthesis, dealloying, electrodeposition, compaction/sintering by combustion, and industrial methods and their key features are discussed. In ***Chapter 3***, the strategies for the functionalization of metal foams to create novel electrode architectures – exploring the advantages of the unique 3D properties of these structures – are summarized. In ***Chapter 4***, a comprehensive review of the key properties of 3D metal foams is provided, including surface area, porosity, roughness, electrical conductivity, thermal conductivity and mechanical stability. A brief description of the techniques used to characterize them is also presented. In ***Chapter 5***, the electrochemical behaviour of metal foam-based electrodes for supercapacitor applications and their most important features are described and analysed. In ***Chapter 6***, an overview of the most recent trends in the application of metallic foams in the development of electrodes for modern batteries is provided. The latest advances are discussed, and their advantages highlighted. In ***Chapter 7***,

Transition Metal-based Nanofoams for Electrochemical Systems: Manufacturing, Characterization and Applications
By Maria de Fátima Montemor, Diana M. Fernandes, Alberto Adan-Mas, Ana Catarina Alves, Gabriel Garcia Carvalho and Inês S. Marques

Published by the Royal Society of Chemistry, www.rsc.org

an overview of the recent progress in metal-based foams for energy-related reactions such as the hydrogen evolution reaction (HER), oxygen evolution reaction (OER) and oxygen reduction reaction (ORR) is provided. The current challenges and future perspectives to further advance the rational design of high-performance transition metal-based foam electrocatalysts are also discussed. In ***Chapter 8***, the current challenges of metal foams, namely, regarding the manufacturing, characterization, and application are summarized and the future directions discussed.

Maria de Fátima Montemor
Diana M. Fernandes
Alberto Adan-Mas
Ana Catarina Alves
Gabriel Garcia Carvalho
Inês S. Marques

Contents

Transition Metal-based Nanofoams for Electrochemical Systems: Manufacturing, Characterization and Applications
By Maria de Fátima Montemor, Diana M. Fernandes, Alberto Adan-Mas, Ana Catarina Alves, Gabriel Garcia Carvalho and Inês S. Marques

Published by the Royal Society of Chemistry, www.rsc.org

CHAPTER 1

Introduction

1.1 Introduction

The urgent need to address the energy crisis worldwide has led to crucial improvements in electrochemical energy conversion and storage devices. Currently, promising devices for flexible and sustainable energy solutions comprise fuel cells that involve electrocatalytic reactions, such as oxygen and hydrogen evolution reactions, and batteries and supercapacitors that involve electrochemical processes. Nowadays, there is a huge demand for more reliable, efficient and less expensive energy storage solutions and, therefore, materials tailored for the specificities of these devices have become a crucial area of work. However, the development of such materials faces many challenges, subsequently pressing scientists to develop more advanced solutions based on cutting-edge functional materials.

To date, in the domain of fuel cells, noble metal-containing electrocatalysts (Pt, RuO_2 and IrO_2) have been considered the most active and effective materials for electrocatalytic reactions, capable of lowering overpotential values to reach higher energy conversion efficiency.[1] Nevertheless, the scarcity, geopolitical dependence, high cost and sometimes toxicity and low stability of these noble metal-based catalysts are inhibiting their application on a larger scale.[1,2] Also, researchers in Europe are looking for new families of materials that are easy to produce and that can leverage the independence of the EU industry from critical raw materials and geopolitical constraints. Presently, there is an enormous interest in finding alternative noble metal-free materials to be employed in the electrocatalytic processes in fuel cells.

Concerning electrochemical energy storage systems, there is a countless portfolio of materials that can be used as electrode materials. In modern

Transition Metal-based Nanofoams for Electrochemical Systems: Manufacturing, Characterization and Applications
By Maria de Fátima Montemor, Diana M. Fernandes, Alberto Adan-Mas, Ana Catarina Alves, Gabriel Garcia Carvalho and Inês S. Marques

Published by the Royal Society of Chemistry, www.rsc.org

lithium-ion (Li-ion) batteries, various lithium-containing metal oxides have been widely used, for instance, materials based on cobalt, manganese and nickel (or combinations thereof). Despite being a relatively well-established technology, new compositions, especially cobalt free, are continuously under development, aiming for higher capacity, longer stability and enhanced cycling performance. Other emerging Li-free technologies, such as metal–air and sodium-ion (Na-ion) batteries, also require stable and highly effective electrode materials, some of them with radically new compositions and properties. Currently, all these technologies are evolving extremely quickly, being accelerated by the huge industrial demand and the many break-through achievements being reported.

Besides batteries, supercapacitors are another category of energy storage devices that are becoming increasingly important. Carbon and transition metal oxides (MnO_2, RuO_2) have been widely reported as electrode materials for ultracapacitors and pseudocapacitors, respectively.[3,4] Although pseudocapacitors achieve higher capacity, due to the contribution of faradaic processes, they may present resistive features and slower ion diffusion rates that limit the high-power performance of the devices compared to conventional double-layer ultracapacitors. Consequently, nowadays, new material compositions, additives, electrolytes and electrode architectures are intensively pursued for supercapacitor-related devices[4,5] and for modern batteries, including post-lithium technologies.[6,7] Overall, the need for more efficient and stable materials is at the vortex of the current developments in the field of electrochemical energy conversion and storage. These technologies cannot evolve unless more reliable and tailored electrode materials are developed and introduced in the electrodes of these devices. The future of electrochemical energy storage is completely dependent on more advanced materials and surface science and engineering. This pushes scientists into new fabrication routes and functionalization strategies in close interdependence with the requirements imposed by the industry for the uptake of new material families.

Amongst the different strategies to fine-tune the properties of electrode materials, optimizing the electrochemical active surface area has been paramount for developing more efficient electrodes. For that reason, highly porous materials have been extensively investigated as electrode materials for electrochemical applications. These porous materials, with a variety of different chemical compositions, possess important advantages such as large specific surface area and low density with tailored pore size, connectivity and pore distribution. Within this classification, metal foams are described as three-dimensional (3D) frameworks with interconnected porous structures (nano-, meso- and macropores) that combine the intrinsic physical and mechanical properties of metals with a highly porous nanoarchitecture.[8] These peculiar nanostructures provide increased surface area, high specific strength, decreased relative density (ρ_{foam}/ρ_{bulk}), a rich pore structure and superior electrical conductivity,[8] which are features required for several applications, particularly in energy conversion and storage electrodes. Such properties place metal foams in a very attractive and unique category, as evidenced in Figure 1.1.

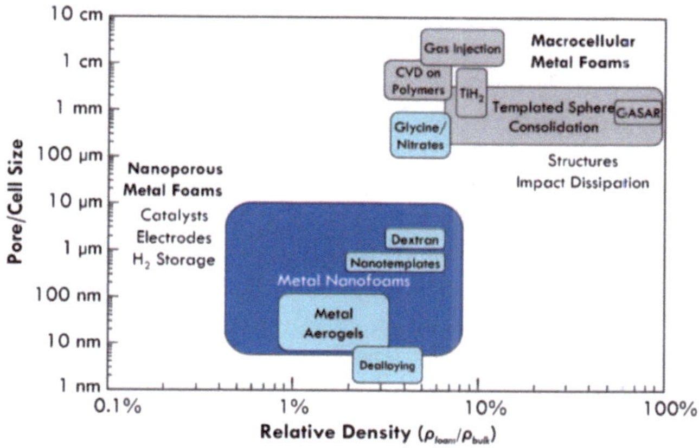

Figure 1.1 Pore/cell size *versus* relative density plot for porous metallic materials. Reproduced from ref. 8 with permission from John Wiley and Sons, Copyright 2010.

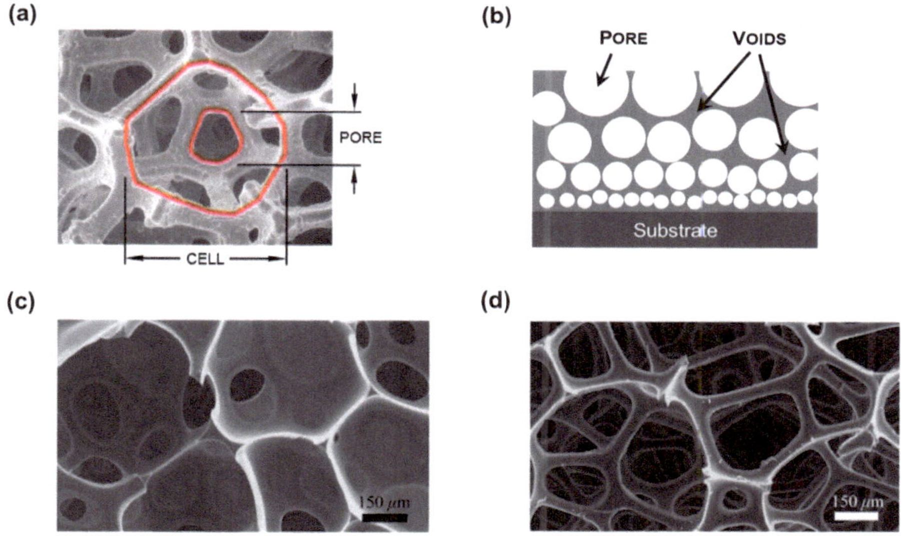

Figure 1.2 Images and nomenclature of the structural elements of a cellular metal: (a) definition of cell foam pore and cell size. Reproduced from ref. 9 with permission from NASA and the authors. (b) Schematic illustration of the definition of cell, pore and void. Example of a (c) closed-cell foam and (d) open-cell foam. Reproduced from ref. 10 with permission from Elsevier, Copyright 2006.

Foams are composed of cells, pores and voids, as shown in Figure 1.2(a). Cells are defined as a multi-faceted polyhedral solid, while pores are the windows that appear between foam ligaments.[9,10] In general, foams are classified as open- or closed-cell according to their cell structure organization. For instance, open-cell foams, as depicted in Figure 1.2(a), exhibit interconnected pore networks, while closed-cell foams show a sealed pore structure, as demonstrated in Figure 1.2(b). Foams may exhibit both pores and voids, which differ mainly in formation and size. On the one hand, pores are open regions formed near the surface, for example, by trapped gases that produce nano- and mesopores. On the other hand, voids are defined as the accumulation of vacancies in the bulk material that produce an empty space from meso- to macrostructures.

The design and preparation of 3D metal micro- and nanofoams underwent significant evolution over the last decade and, consequently, great progress has been made in the synthesis of transition metal-based foams. For example, depending on the method and experimental conditions of the synthesis, the features and key properties of the metal foams may change significantly, determining the metrics and electrochemical properties of the electrode materials in the final application. Simultaneously, it has been possible to achieve a better understanding of the correlation between the composition of the foam and its structural and physical/chemical properties, and their final electrochemical performance as electrode material. According to the literature, several parameters can affect the performance of the foam material in its final electrochemical application. In particular, the following factors must be highlighted:[11–17]

1. the chemical composition of the foam;
2. the crystal structure and crystalline facets;
3. morphology and roughness;
4. surface area and pore size and pore distribution;
5. thermal and chemical stability;
6. electronic conductivity;
7. electrochemical activity;
8. mechanical stability;
9. cost.

The literature provides excellent review articles introducing micro- and nanofoams for numerous applications.[2,18–28] Nonetheless, considering the relevance that nanoporous foams are gaining, it is important to have a more comprehensive overview of the state-of-the-art progress of metal foams, especially concerning synthesis, properties, functionalization and applications for electrochemical energy storage and conversion. Therefore, this book provides an overview of the key aspects of porous metal nanofoams, discusses different synthesis routes and highlights the correlation between these routes and the required properties for higher electrocatalytic and electrochemical performance in different applications. In this way, this book fills a gap that exists in the current literature by evidencing and discussing the potential of 3D metal foams as a building block for the design of novel electrode architectures for electrochemical devices, with a particular focus on their electrocatalytic performance and application in supercapacitors, batteries and energy-related electrocatalysis.

References

1. H. Tao, Y. Gao, N. Talreja, F. Guo, J. Texter, C. Yan and Z. Sun, *J. Mater. Chem. A*, 2017, **5**, 7257–7284.
2. W. Zhu, R. Zhang, F. Qu, A. M. Asiri and X. Sun, *ChemCatChem*, 2017, **9**, 1721–1743.
3. Z. Zhao, T. Shen, Z. Liu, Q. Zhong and Y. Qin, *J. Alloys Compd.*, 2020, **812**, 152124.
4. C. Du, E. Han, L. Sun, S. Qiao and L. Li, *Ionics*, 2019, **26**, 383–391.
5. Y. Liu, W. Li, X. Chang, H. Chen, X. Zheng, J. Bai and Z. Ren, *J. Colloid Interface Sci.*, 2020, **562**, 483–492.
6. J. Zhang, R. Chu, Y. Chen, H. Jiang, Y. Zhang, N. M. Huang and H. Guo, *Ceram. Int.*, 2018, **44**, 16219–16226.
7. C.-s. An, B. Zhang, L.-b. Tang, B. Xiao, Z.-j. He and J.-c. Zheng, *Ceram. Int.*, 2019, **45**, 13144–13149.
8. B. C. Tappan, S. A. Steiner 3rd and E. P. Luther, *Angew. Chem., Int. Ed.*, 2010, **49**, 4544–4565.
9. S. Ryan, T. Hedman and E. Christiansen, *NASA Technical Reports*, 2009, JSC-CN-18720.
10. Y.-J. Lee and K. B. Yoon, *Microporous Mesoporous Mater.*, 2006, **88**, 176–186.
11. K. I. Siwek, S. Eugénio, D. M. F. Santos, M. T. Silva and M. F. Montemor, *Int. J. Hydrogen Energy*, 2019, **44**, 1701–1709.
12. K. I. Siwek, S. Eugénio, T. M. Silva and M. Fatima Montemor, *Energy Technol.*, 2019, **7**, 1801139.
13. X. Ma, Y. Chang, Z. Zhang and J. Tang, *J. Mater. Chem. A*, 2018, **6**, 2100–2106.
14. J. Cao, J. Zhou, Y. Zhang, Y. Wang and X. Liu, *ACS Appl. Mater. Interfaces*, 2018, **10**, 1752–1760.
15. J. Li, G. Wei, Y. Zhu, Y. Xi, X. Pan, Y. Ji, I. V. Zatovsky and W. Han, *J. Mater. Chem. A*, 2017, **5**, 14828–14837.
16. M. Yan, Y. Yao, J. Wen, L. Long, M. Kong, G. Zhang, X. Liao, G. Yin and Z. Huang, *ACS Appl. Mater. Interfaces*, 2016, **8**, 24525–24535.
17. H. Zhou, Y. Wang, R. He, F. Yu, J. Sun, F. Wang, Y. Lan, Z. Ren and S. Chen, *Nano Energy*, 2016, **20**, 29–36.
18. X. Wu, G. He and Y. Ding, *Electrochem. Energy Rev.*, 2020, **3**, 541–580.
19. H. J. Qiu, H. T. Xu, L. Liu and Y. Wang, *Nanoscale*, 2015, **7**, 386–400.
20. I. McCue, E. Benn, B. Gaskey and J. Erlebacher, *Annu. Rev. Mater. Res.*, 2016, **46**, 263–286.
21. T. Song, M. Yan and M. Qian, *Corros. Sci.*, 2018, **134**, 78–98.
22. I. Duarte and J. M. F. Ferreira, *Materials*, 2016, **9**, 79.
23. T. Juarez, J. Biener, J. Weissmüller and A. M. Hodge, *Adv. Eng. Mater.*, 2017, **19**, 1700389.
24. A. Varma, A. S. Mukasyan, A. S. Rogachev and K. V. Manukyan, *Chem. Rev.*, 2016, **116**, 14493–14586.

25. L. F. Dumée, L. He, B. Lin, F.-M. Ailloux, J.-B. Lemoine, L. Velleman, F. She, M. C. Duke, J. D. Orbell, G. Erskine, P. D. Hodgson, S. Gray and L. Kong, *J. Mater. Chem. A*, 2013, **1**, 15185–15206.
26. T. A. Rebbecchi and Y. Chen, *J. Mater. Res.*, 2017, **33**, 2–15.
27. B. J. Plowman, L. A. Jones and S. K. Bhargava, *Chem. Commun.*, 2015, **51**, 4331–4346.
28. R. Du, X. Jin, R. Hübner, X. Fan, Y. Hu and A. Eychmüller, *Adv. Energy Mater.*, 2019, **10**, 1901945.

CHAPTER 2

Methodologies to Prepare Metal Foams

2.1 Templating Method

Templating is extensively used to fabricate porous metallic materials with finely controlled morphologies. The process makes use of a template that is removed by chemical or electrochemical leaching, producing a replica of the sacrificial structure. The general process to obtain foams by templating is depicted in Figure 2.1 and it comprises three steps: (1) the synthesis of the template; (2) its metallization and (3) the removal of the sacrificial template.[1] The pore dimension and morphology will be determined by both the template structure and the metallization technique used to deposit the metal onto the surface of the template. In this chapter, we will focus on chemical/electrochemical dealloying and dynamic hydrogen bubbling as templating techniques. Other templating methods can also be found in the literature, particularly, sacrificial templates such as soft templating and hard (inorganic) templating.[1–4] Templating has important advantages, including the possibility of tailoring the original template to design different porous structures. However, the process also requires multiple steps, including an aggressive template removal with specific chemical baths as the final step. Therefore, it is critical to develop simpler, cost-effective and environmentally-friendly routes to produce these materials.

Transition Metal-based Nanofoams for Electrochemical Systems: Manufacturing, Characterization and Applications
By Maria de Fátima Montemor, Diana M. Fernandes, Alberto Adan-Mas, Ana Catarina Alves, Gabriel Garcia Carvalho and Inês S. Marques

Published by the Royal Society of Chemistry, www.rsc.org

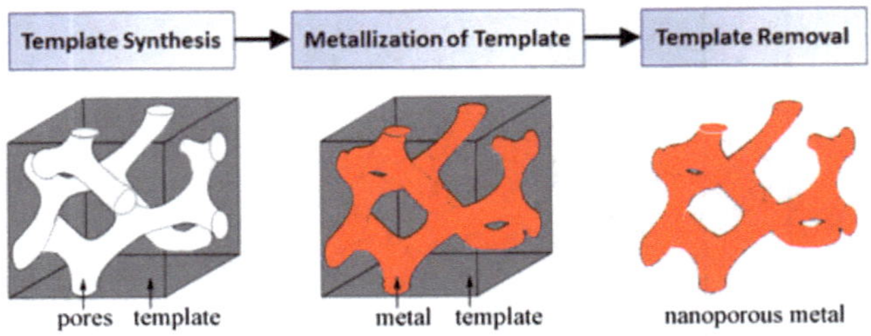

Figure 2.1 General process followed during template-based synthesis of nanoporous metals. Reproduced from ref. 1 with permission from Springer Nature, Copyright 2018.

2.1.1 Dealloying

Based on a galvanic corrosion process, dealloying is used in the synthesis of porous metal foams by selectively leaching electroactive metals from an alloy to create nanopores (15 to 150 nm). Considering an alloy composed of two metals, a requisite is that one metal shall be nobler than the other, thus presenting more resistance to chemical or electrochemical leaching.[5] The mechanism follows the leaching of the less noble metal of the alloy (sacrificial template) by surface diffusion, resulting in an open porous network with large surface area (Figure 2.2).[6] The larger the difference between the electrochemical potentials of the metals involved, the simpler and more efficient the dealloying process. The main advantage of this process is that the final nanostructure of the porous material can be effectively tailored by controlling the alloy composition (including the electrochemical potential difference between the metals used) and the leaching parameters applied during the dealloying process.

Dealloying is subsequently categorized depending on the leaching method, in either chemical or electrochemical dealloying. In the former, the removal of the alloy's components is achieved by dissolving one or more phases in an acidic or alkaline aqueous solution. The evolution of the pore system created during chemical leaching is controlled by experimental factors such as a foam precursor, composition of the solution and time of leaching. These parameters affect the leaching rate of the less noble metal as well as the rate of surface diffusivity and capability of rearrangement of the more noble metal.[7] To achieve effective chemical dealloying, some aggressive inorganic acids might be required. This leads to a manufacturing process that may not be so environmentally friendly. Therefore, the search for newer and more environmentally compliant dealloying routes is still an emergent research topic.

Electrochemical dealloying is based on the selective dissolution of the more electrochemically active material by applying either (i) a potential difference (by cyclic voltammetry or pulsed voltage mode), or (ii) a constant

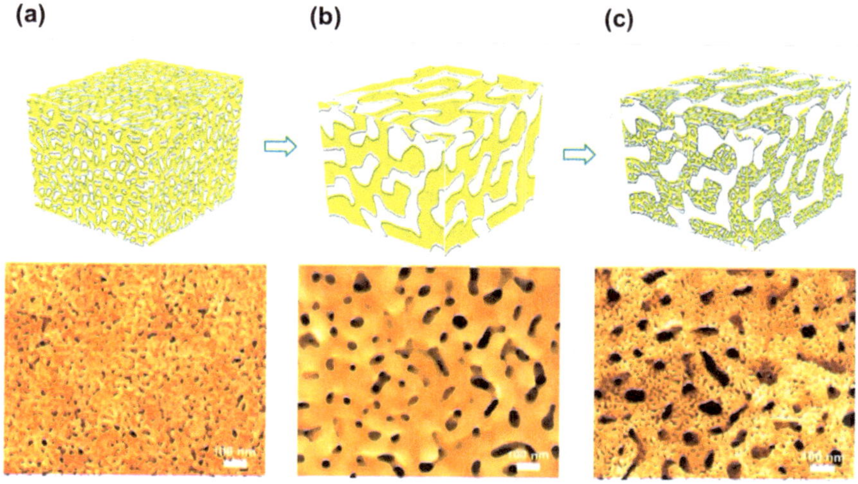

Figure 2.2 Scheme of the fabrication of 3D hierarchical nanoporous gold (NPG) by dealloying alongside the corresponding SEM image of (a) Au–Ag alloy (average nanopore size of 20 nm); (b) nanoporous Au–Ag alloy (pore size of ~80–100 nm); and (c) nanoporous gold (pore size of 5–20 nm on the 80–100 nm ligaments). Reproduced from ref. 6 with permission from Springer Nature, Copyright 2016.

potential (potentiostatic mode). The advantage of electrochemical dealloying is its precision in the steering process, which allows control of the partial dissolution and surface reorganization of the nanoparticles and enables a finer tuneability over the composition and porosity of the resulting metallic foam.[8,9] A rational selection of the metals used may enable an efficient dealloying process and successful recovery of the less noble metal (the one dissolved). However, control of the process and the purity and composition of the final foam obtained is a major research topic that still requires important advances. As such, this is an area of work that currently captures the interest of the scientific community.

Selective dealloying is applied mostly to noble metal-containing alloys, such as Au–Ag,[6,10–12] Ag–Al[13–15] and Au–Al.[16–18] Consequently, an extensive amount of research work has been developed owing to the good corrosion resistance of noble metals. However, in the view of industrial energetic applications, the prohibitive cost of noble metal alloys limits their use in large-scale production. Subsequently, the fabrication of porous materials from cheaper transition metal-based alloys has been intensively investigated.[19] Among others, porous copper foams produced by dealloying Cu–Zn,[20–24] Cu–Ti,[7,25] Ni–Cu[9] and Cu–Al alloys[19,26–33] have been prepared for assembling electrodes for different energy applications. Hereafter, we will discuss the main features and results of chemical and electrochemical dealloying of Ni-, Cu-, Al- and Zn-based foams.

2.1.1.1 Chemical Dealloying

Al–Cu alloys are one of the most studied systems to produce foams by chemical dealloying, owing to the ease of etching in alkaline and acidic media. For that reason, the discussion on chemical dealloying will be essentially focused on these alloys. The ratio between the two components on the alloys is essential to develop an optimized porous structure. Liu *et al.*[19] have reported the fabrication of nanoporous copper (NPC) foams using Al–Cu alloys with 33 to 50 at% of Cu under alkaline chemical leaching (10 wt% NaOH aqueous solution at room temperature). In their work, XRD analysis showed the presence of three phases in the Al–Cu alloys: Al_2Cu, AlCu and face-centred cubic (fcc) Cu phases, depending on the stage of dealloying and the content of copper in the alloy (Figure 2.3). For instance, before the dealloying process shown in Figure 2.3(a), the 33% Cu alloy (Al-33 Cu) exhibited a single Al_2Cu phase, while 35, 40 and 50% Cu alloys (Al-35, 40 and 50 Cu, respectively) were composed by mixed Al_2Cu and AlCu phases. After dealloying, as depicted in Figure 2.3(b), only the fcc Cu phase was observed in the XRD pattern of Al-33 and 35 Cu alloys, while Al-40 and 50 Cu alloys showed mixed phases, and a lower amount of fcc Cu was detected.

These XRD observations were supported by SEM imaging after dealloying, demonstrating a bi-continuous interpenetrating ligament pore structure for Al-33 and 35 Cu alloys with length scales of 20–25 and 25–30 nm, respectively.[19] The SEM images are shown in Figure 2.3(c)–(f). For higher Cu content alloys,

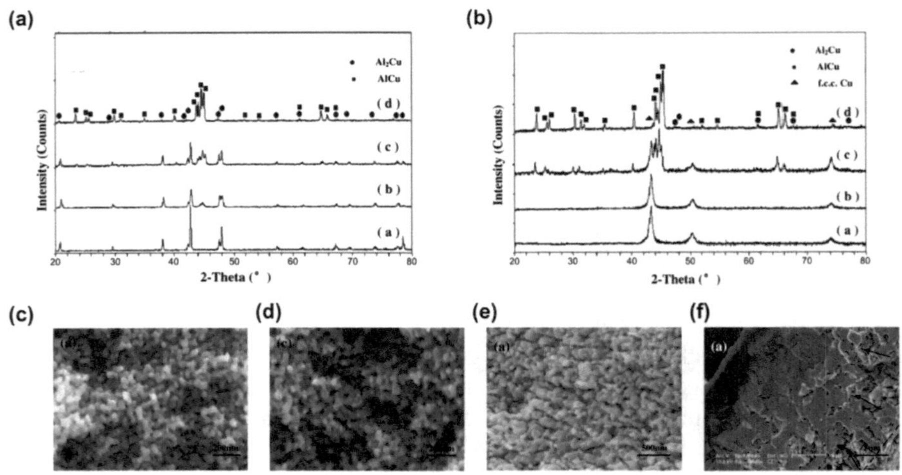

Figure 2.3 XRD patterns of Al–Cu alloys (a) before and (b) after dealloying. Inset caption of (a) Al-33 Cu, (b) Al-35 Cu, (c) Al-40 Cu and (d) Al-50 Cu. SEM images of NPC by chemical dealloying of (c) Al-33 Cu (ligament pore-structure of 20–25 nm), (d) Al-35 Cu (25–30 nm), (e) Al-40 Cu (60–80 nm) and (f) Al-50 Cu (150–200 nm) alloys, in a 10 wt% NaOH aqueous solution. Reproduced from ref. 19 with permission from Elsevier, Copyright 2011.

such as Al-40 Cu, SEM images showed microstructures with a uniform porous structure, see Figure 2.3(e), and EDS analysis detected residual Al, which was indicative of un-dealloyed AlCu in the network. The dealloying of an Al-50 Cu sample induced the formation of a non-uniform porous structure, and constriction of the percolating paths in the inner space of the sample. Such observations suggested that the elimination process of Al only occurred at the surface layers of the alloy.[19] In a dual-phase alloy, Al_2Cu shows higher electrochemical activity compared to AlCu, thus, Al_2Cu is preferentially dissolved, forming percolating paths at the inner structure of the alloy.[19] Hence, Al-33 and 35 Cu alloys were completely dealloyed in the alkaline solution, forming a more uniform NPC structure compared to Al–Cu alloys with higher Cu content. A synergistic effect was observed when both phases were present.

The synergistic dealloying of Al_2Cu and Al–Cu phases was also observed in acidic solutions, with a change in corrosion behaviour of the constituents of the precursor alloy. For example, using an acidic aqueous solution of 5 wt% HCl in the Al–Cu alloy, XRD indicated the formation of a single-body centred tetragonal Al_2Cu and fcc Cu after dealloying the Al-15 Cu alloy.[31,34] These results are related to the good stability of AlCu in HCl solution, that stabilizes a mixture of Al_2Cu and α-Al nanoporous copper foam.[30–32,34] Similar results were reported by several authors[26–28,32,33] for both this alloy and Mg–Cu, reported by Liu *et al.*[34]

Cu–Zn alloys are also used as precursor material for the synthesis of copper nanostructures, as described by Mao *et al.*[20] The addition of complexing agents allowed the formation of coarsening ligaments on the surface due to the rapid surface diffusion of Cu induced by 5 wt% HCl. The structure $Zn_{70}Cu_{30}$ was found to be an appropriate precursor for the synthesis of nanoporous Cu structures because of its pore morphology, uniform ligaments and low Zn residue.[20] Despite the advantages of this route, sometimes harsh chemicals might be required making these processes not totally environmentally compliant. Also, the number of metal combinations is limited and mainly restricted to binary alloys.

2.1.1.2 Electrochemical Dealloying

An alternative to chemical dealloying is electrochemical dealloying, where the selective leaching of a metal is achieved by applying current. Based on the redox reactions of various metals, it is possible to understand the dealloying process of Ni–Cu, Cu–Zn and Cu–Al. The following equations represent the half reactions of different metals including the respective electrochemical potential *vs.* standard hydrogen electrode (SHE):[35]

$$Ni^{2+} + 2e^- \leftrightarrow Ni\left(-0.26\text{V } vs.\,\text{SHE}\right) \tag{2.1}$$

$$Zn^{2+} + 2e^- \leftrightarrow Zn\left(-0.76\text{V } vs.\,\text{SHE}\right) \tag{2.2}$$

$$Al^{3+} + 3e^- \leftrightarrow Al\left(-1.66\text{V } vs.\,\text{SHE}\right) \tag{2.3}$$

$$Cu^{2+} + 2e^- \leftrightarrow Cu\left(+0.34\text{V } vs.\,\text{SHE}\right) \tag{2.4}$$

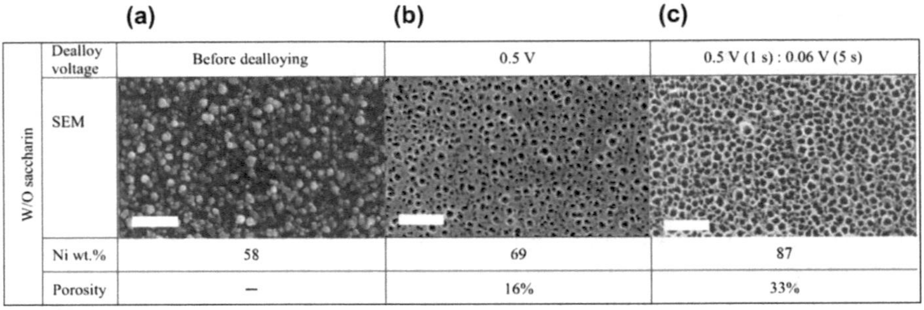

Figure 2.4 SEM images, Ni content in wt% and porosity of the as-deposited and as-dealloyed Ni–Cu films by pulsed voltage waveforms and constant voltage. Reproduced from ref. 9 with permission from Elsevier, Copyright 2014.

Zhang *et al.*[9] reported the synthesis of Ni porous structures by electrodepositing Ni and Cu on indium tin oxide-coated glass that was further dealloyed by applying either pulsed voltage waveforms (V_1 = 0.5 V for 1 s and V_2 = 0.06 V for 5 s) or constant voltage (ConstV) of 0.5 V for 30 minutes (Figure 2.4). Additives in the electroplating bath can also be used to modify the final morphology of the foam. For example, the same work also reported the addition of a surfactant, saccharin, in the electroplating bath that acts as morphology-inductive agent leading to a smoother surface. The Ni porous structures obtained by pulse-dealloyed samples showed larger pores (300–500 nm) and enhanced porosity and thinner ligaments (50 nm) compared to the samples prepared at constant voltage. Moreover, pulse dealloying generated porous Ni films with remarkably higher purities than potentiostatic dealloying (87 *vs.* 69 wt% of Ni).[9] The impact of time during the voltage step (t_1 at 0.5 V and t_2 at 0.06 V) was also investigated in this work:[9] $t_1 = t_2$ = 5 s and t_1 = 1 s, t_2 = 15 s. A similar morphology of the porous Ni film was obtained for smaller and longer pulsing steps, exhibiting smaller porosity (~27%) and lower Ni purity (~80 wt%) than the foam obtained at t_1 = 1 s, t_2 = 5 s. Moreover, by increasing the overall dealloying time to 43 and 180 min, a more compact Ni foam was obtained, with smaller pores.[9] Sun *et al.*[36] studied the formation of nanoporous Ni by electrochemically leaching copper from a $Ni_{0.7}Cu_{0.3}$ alloy which resulted in the production of an interconnected pore structure, with pores ranging from 100–200 nm in diameter. By increasing the percentage of copper in the alloy (up to x = 0.5, in Ni_xCu_{x-1}), an irregular morphology was obtained, which is indicative of the passivation and low mobility of Ni.[36] Nanoporous copper structures have also been fabricated employing electrochemical dealloying by Zhao *et al.*[37] and Song *et al.*[38] Copper nanostructures using Mg–Cu and Al–Cu alloys were obtained, respectively. Such investigations describe the importance of the alloy composition and phase constitution of the electrochemical system. In a similar way, other nanoporous materials have been prepared by electrochemical leaching, for instance, Al-based[39] and Ni-NiO alloys.[40]

Dealloying processes enable the production of nano- to micropores (10 nm to 100 μm) and the synthesis of shaped porous products, mainly ribbons, rods and strips. However, to date, the structural defects by selective alloy corrosion have received little attention and, thus, have been scarcely reported.[41] Moreover, diffusion-limited processes restrain pore and film dimensions, and the applicability of this method is largely limited to noble metal alloys, as aforementioned,[1] which excludes the application of this method on the synthesis of transition metal-based nanofoams, hindering its scale-up and further industrial potential. Although simpler to achieve, the process is more efficient and easier to control when limited to bimetallic alloys. Table 2.1 summarises the experimental conditions for the dealloying of metal alloys reported in literature.

2.1.2 Electrochemical Deposition Under the Dynamic Hydrogen Bubble Template

The dynamic hydrogen bubble template (DHBT) is an electrodeposition technique that uses hydrogen bubbles as a negative dynamic template for the formation of a porous metal structure. In this electrochemical process, a current/voltage is applied to promote the hydrogen evolution reaction (HER) on top of a conductive substrate in the presence of a solution containing metallic ions of interest. The HER is triggered by cathodic overpotentials, reducing H^+ ions to H_2 while producing hydrogen bubbles that act as a dynamic template. The following chemical reactions describe the HER process:

$$2H^+\,(aq) + 2e^- \leftrightarrow H_2\,(g) \tag{2.5}$$

$$H^+\,(aq) + e^- \leftrightarrow H\,(ads) \tag{2.6}$$

$$2H\,(ads) \leftrightarrow H_2\,(g) \tag{2.7}$$

$$H^+\,(aq) + H\,(ads) + e^- \leftrightarrow H_2\,(g) \tag{2.8}$$

The hydrogen bubbles produced by the HER are adsorbed to the surface of the substrate, which disrupts the growth of the metal structure that is, therefore, electrodeposited around the bubbles or within the interstitial spaces between them. The bubble release is affected by the ability of bubbles to coalesce, which is, in turn, dependent on the surface energy of the substrate used.[42] Hence, the formation of the opened porous metal structure of the foam results from the competition between the properties of bubbles by HER and the reduction of metal ions on the substrate's surface. The schematic representation of this process is depicted in Figure 2.5. This electrochemical route results in a nanoporous coating with a high specific surface area that is essential for an efficient mass transport and electrolyte diffusion in many electrochemical processes. Compared with other conventional methods used to prepare foams, DHBT shows high efficiency, versatility, controllability, simplicity and low cost,[43–45] creating porous morphologies without the use of inorganic templates or multi-step procedures.

Table 2.1 Summary of some key electrodeposition parameters to prepare foams by chemical and electrochemical dealloying.

Material	Alloy	Alloy synthesis	Dealloying conditions	Time of reaction	References
Nanoporous Cu ribbons	Al–Cu alloy 33, 35, 40, 50 at% Cu	Voltaic arc heating, in Ar atm; high-frequency induction heating	10 wt% NaOH at RT	8 to 48 h	19
Nanoporous Cu ribbons	Al–Cu alloy 15 at% Cu	Voltaic arc heating, in Ar atm; high-frequency induction heating	5 wt% HCl at 75 °C 10 wt% NaOH at RT	Until no bubbles emerged	31
Nanoporous Cu ribbons	Mg–Cu alloy with 15.5 at% Cu Al–Cu alloy with 15 at% Cu	Voltaic arc heating, in Ar atm; high-frequency induction heating	1 or 5 wt% HCl at RT	3, 5 and 15 min	34
Cu nanoporous foam	Powder mixing of Al, Al_2O_3, NH_4Cl and Cu	Power mixing; packing and sealing; pack-cementation	HCl at 45 °C (pH 0.25)	24 h	26
Nanoporous Cu foam	Al_2Cu	Ball-milling; hot-rolling and -pressing	20 wt% NaOH	1, 2, 3, 5 and 9 h	27
Nanoscale copper foams	Powder mixing Al and Cu 30 at% Cu	Mixing; pressing: sintering (450 °C, 24 h)	5 wt% NaOH	5 h	28
Nanoporous Cu ribbons	Al–Cu alloy 32 at% Cu	Voltaic arc heating, in Ar atm; high-frequency induction heating	5 wt% HCl at 75 °C 10 wt% NaOH at 15 °C	Until no bubbles emerged	32
Nanoporous copper ribbons	Al–Cu alloy 33, 35, 40, 50 at% Cu	High-frequency induction heating; melting-spun	5 wt% HCl at 90 °C	2 to 4 h	33
Nanoporous copper	Mg–Cu alloy 33, 40, 50, 60, 67 at% Cu	Melt-spinning process	0.2 M NaCl	Not given	37
Nanoporous copper	Zn–Cu alloy 20, 30, 40, 60, 50 at% Cu	Melting process, under N_2 atm (900 °C, 2 h)	HCl or NH_4Cl at 70 °C	Until no bubbles emerged	20
Porous copper	Al–Cu alloy 25 at% Cu	Melting under Ar atm; annealing under Ar atm (490 °C, 24 h)	1 M NaCl	Not given	38
Ni foam	Ni–Cu alloy	Electrodeposition at −0.820 V *vs.* SCE for 130 min	Pulse dealloying 0.5 V (1 s); 0.06 V (5 s)	30 min	9
Nanoporous nickel	Ni–Cu alloy	Electrodeposition between −0.75 and −1.0 V *vs.* Ag/AgCl	Not given	Not given	36
Nanoporous metals	Al–M, M = Au, Ag, Pd, Cu	High-frequency induction heating; melting-spun	Different dealloying potentials, 1 M NaCl	30 min	39
Ni-NiO foam	Ni–Cu alloy	Electrodeposition at 2 A cm^{-2} for 180 s	0.5 V, in 1 M H_2SO_4	Not given	40

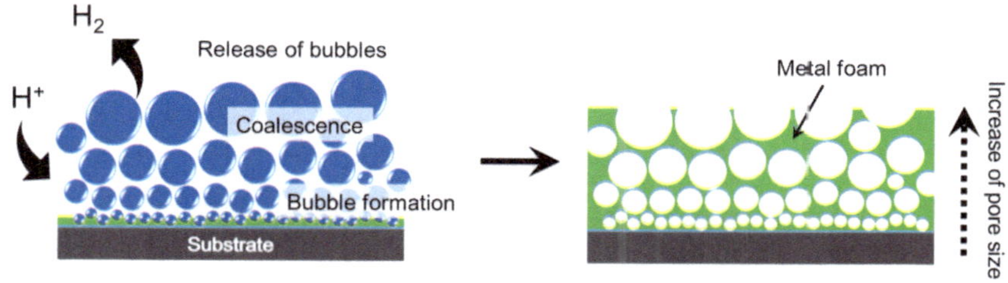

Figure 2.5 Schematic view for the formation of a metal foam by cathodic electrodeposition using the dynamic hydrogen bubble template.

The preparation of metal nanofoams by cathodic electrodeposition has been a hotspot of several works, particularly those oriented to electrocatalysis and energy storage[43–65]. The nanostructure features of porous metallic materials are better understood by the investigation of bubble behaviour, nucleation of the metal ions and the electrochemical growth mechanisms. By controlling the experimental parameters of electrodeposition, particularly the current density and time of deposition, and the composition of the solution (additives), metallic foam structures can be obtained with tailored morphology and porosity, pore dimension and foam composition. Furthermore, by using different metallic precursors in the original bath, multimetal foams can be easily produced, and different alloy compositions can be designed. Table 2.2 summarizes the experimental electrodeposition conditions used to prepare metal foams by DHBT.

2.1.2.1 *Applied Current Density and Time of the Electrodeposition*

The applied current density (or overpotential) has a significant influence on the formation of 3D foam structures. The hydrodynamic conditions near the electrode surface changes during electrodeposition depending on the extension of the hydrogen evolution process and diffusion layer, varying the rate of deposition of metal ions onto the surface.[47,53] For instance, Lange *et al.*[48] reported an investigation on Cu–Fe foams synthesized by applying current densities between 0.5 and 1.8 A cm^{-2}. With a deposition time of 90 s, the formation of dendrites on the granular film was observed and, with increasing current density, a larger ramified and dense structure was developed, as shown in Figure 2.6(a)–(d).[48] Similar results were presented by Cardoso *et al.*,[55] in which the increased current density induced the formation of a compact wall structure of Ni–Cu metallic foams with dendrites, with the decrease of pore density and enhanced heterogeneity of the pore size distribution (Figure 2.6(e)–(f)). The authors in ref. 55 stated that the larger amount of metal deposited at higher currents resulted in a different porosity of the two foams. Comparable results were reported by other authors.[50,60–62]

Table 2.2 Summary of key electrodeposition parameters to prepare foams by DHBT.

Material	Electrodeposition conditions: Substrate	Current applied/A cm^{-2}	Time/s	Electrolyte	References
Cu–M metal foams (M = Fe, Co, Ni)	AISI 304 stainless steel (1 cm^2)	1.8	180	0.01 M $CuSO_4·5H_2O$, 1.5 M H_2SO_4, 1 M HCl and $Fe(H_2NO_3S)_2$ or 0.5 M $CoSO_4·7H_2O$ or $NiSO_4·6H_2O$	56
Ni–Co foams	AISI 304 stainless steel (1.65 cm^2)	1	150	0.1 M $NiCl_2$, 0.1 M $CoCl_2$, 1 M NH_4Cl, 1 M NaCl	47
3D Ni foams	AISI 304 stainless steel (1.65 cm^2)	1	150	0.1 M $NiCl_2·6H_2O$, 1–2 M NH_4Cl, 1 M NaCl	46
Ni–Mn–Sn foams	Copper foil	0.100 to 0.180	Not given	0.03 M $NiSO_4·6H_2O$; 0.59 M $MnSO_4·H_2O$; 0.01 M $SnSO_4$; 1 M $(NH_4)_2SO_4$; 0.01 M EDTA	50
Ni foams	AISI 304 stainless steel (1.65 cm^2)	1	150	0.1 M $NiCl_2·6H_2O$, 1 M NH_4Cl, 1 M NaCl	49
3D Ni foams	AISI 304 stainless steel (0.67 cm^2)	1	120	0.1 M $NiCl_2·6H_2O$, 2 M NH_4Cl	51
Co foams	AISI 304 stainless steel (0.64 cm^2)	1.8	Different times	0.1 M $CoCl_2$, 2 M NH_4Cl, 0.1% m/m agar	58
Cu–Fe foams	AISI 304 stainless steel (1.65 cm^2)	0.5 to 1.8	30–180	0.5 M $Fe(H_2NO_3S)_2$, 0.01 M $CuSO_4$, 1.5 M H_2SO_4, 1 M HCl	48
Cu films	Copper foil	0.2 to 2.0	5–50	0.1 M $CuSO_4$, 0.5 M H_2SO_4, $(NH_4)_2SO_4$, Na_2SO_4, NaCl and CTAB	43
Cu foams	Copper foil	3	4	0.2 M $CuSO_4$, 0.7 M H_2SO_4, 1.2 M $(NH_4)_2SO_4$, 0.4 mM BTA	44
Cu foam	Copper foil	2	30	0.2 M $CuSO_4$, 1.0 M H_2SO_4, 15 mM NaCl	64
Cu foam	Copper foil	0.5 to 2.5	Not given	0.2 M $CuSO_4$ + 1.0 M H_2SO_4 + 2.0 NH_4Cl	45
Cu foam	Copper foil	0.010	300	0.12 M $CuSO_4$, 0.40 M H_2SO_4	52
Ni–Cu foams	AISI 304 stainless steel (1 cm^2)	2 to 3	180	0.5 M $NiSO_4$, 1.5 M H_2SO_4, 1 M HCl, 0.01 M $CuSO_4$	55

Ni–S porous film	Copper foil	3.0	20	1.5 M H_2SO_4, 50 g L^{-1} $CuSO_4$, 0.2 mL L^{-1} HCl, 71 g L^{-1} Na_2SO_4	53
Ni–Sn foams	Pt/Ti/Sn wafer (0.1256 cm^2)	−4 V	300	$NiCl_2 \cdot 6H_2O$, $SnCl_2$, 1 M H_2SO_4	57
Ni–Co–Fe foam	Polycrystalline Ti	1	1800	0.09 M $NiSO_4$, 0.03 M $FeSO_4$, 0.053 M $CoSO_4$, 0.4 M H_3BO_3, 0.28 M NH_4Cl	59
Ni–Cu foam	Stainless steel (1.65 cm^2)	0.6 to 1.8	30–180	0.5 M $NiSO_4 \cdot 7H_2O$, 1.5 M H_2SO_4, 1 M HCl, 0.01 M $CuSO_4 \cdot 5H_2O$	60
Cu foam	AISI 304 stainless steel (0.5 cm^2)	0.5 to 1.0	180–450	0.05 M $CuSO_4$, 0.5 M H_2SO_4	61
Cu foam	Copper foil	2	5–50	0.2 M $CuSO_4$, 0.5 M H_2SO_4, Na_2SO_4 and $(NH_4)_2SO_4$	62
FeNi 3D foams	Carbon steel	1.2 A cm^{-2}	180 s	0.1 M $FeCl_3$; 0.05 M citric acid; 2 M NH_4Cl; $NiCl_2 \cdot H_2O$	65
Ni–Cu foam	AISI 304 stainless steel (1 cm^2)	1 to 1.8	90–180	1.5 M H_2SO_4, 1 M HCl, 0.01 M $CuSO_4 \cdot 5H_2O$, 0.5 M $NiSO_4 \cdot 7H_2O$	63
Ni–Co–Cu foam				1.5 M H_2SO_4, 1 M HCl, 0.01 M $CuSO_4 \cdot 5H_2O$, 0.5 M $NiSO_4 \cdot 7H_2O$, 0.5 M $CoSO_4 \cdot 5H_2O$	

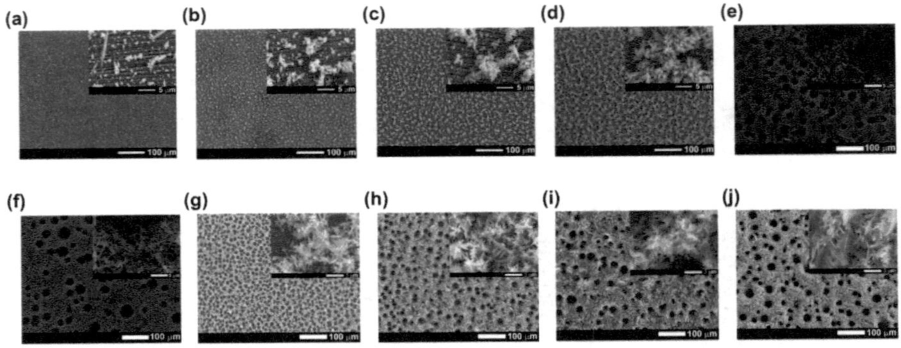

Figure 2.6 SEM images of Cu–Fe foams prepared by DHBT electrodeposited for 90 s at (a) 0.5 and (b) 1, (c) 1.5 (d) 1.8 A cm^{-2}. Reproduced from ref. 48 with permission from Elsevier, Copyright 2015. SEM images of Ni–Cu foams deposited by DHBT for 180 s at (e) 2 and (f) 3 A cm^{-2}. Reproduced from ref. 55 with permission from the Royal Society of Chemistry. SEM images of Ni–Cu foams deposited by DHBT at 1.8 A cm^{-2} for (g) 30, (h) 60, (i) 120 and (j) 180 s. Reproduced from ref. 60 with permission from Springer Nature, Copyright 2014.

The electrodeposition time also has a significant influence on the morphology of the metal foam. For example, both of works reported in ref. 60 and 63 studied the influence of deposition time on the morphology of Ni–Cu metallic foams shown in Figure 2.6(g)–(j). At 1.8 A cm^{-2} for 30 s of deposition, a 3D foam film was obtained with an interconnected dendrite structure on the wall of the pores. With the increase of the deposition time, a branched fern-like structure was formed (Figure 2.6(j)) that induced an increase of the average surface pore area from 108 μm^2 (deposition at 60 s) to 225 μm^2 (deposition of 180 s). This enhancement of the average surface pore area of the metal foam was related to the increased empty space in the pore walls due to the formation of an open dendritic structure. These results are in accordance with others reported elsewhere on Ni–Cu.[48]

2.1.2.2 *Solution Composition: Addition of Inorganic Agents*

The composition of the electrolyte used in a DHBT experiment determines the micro- and nanoscopic morphology and crystallinity of the metal foams. Usually, the DHBT solution is composed by a H^+-containing precursor capable of releasing hydrogen bubbles.[43–45,48,52,53,55,60] Additives, complexing agents and surfactants added to the electrolyte solution are a key factor that critically affect bubble behaviour and contribute significantly to the crystallization process. Such features influence the dimension, shape and density of the pores in the metal foam. The difference in the overall morphology results from a change in the surface tension and viscosity of the electrolyte with additives.[64] Examples of inorganic and organic additives that have been added to the electrolyte are NH_4Cl,[43,46–51,53,55–58,64] benzotriazole (BTA),[44] cetrimonium bromide (CTAB),[43] polyethylene glycol (PEG), agar-agar[58] and ethylenediamine tetraacetic acid (EDTA).[50,56]

The use of inorganic salts as a morphological additive induces the growth of metal ion deposit and improves the electrolyte conductivity.[46] Particularly, aqueous NH_4Cl may be one of the most used additives in the supporting electrolyte for the electrodeposition of porous metal coatings,[43,46–51,53,55–58,64] contributing to the release of hydrogen bubbles according to eqn (2.9):

$$2NH_4^+ + 2e^- \leftrightarrow H_2 + 2NH_3 \tag{2.9}$$

Siwek *et al.*[46] reported the preparation of 3D Ni foams in different electrolytes, investigating the effect of the concentration of NH_4Cl on the morphology (including shape and texture) and the structure growth of the foam. For instance, the increase of NH_4Cl concentration from 1 M (foam 2 and 3 in Figure 2.7) to 2 M (foam 1 in Figure 2.7) produced Ni foams with higher

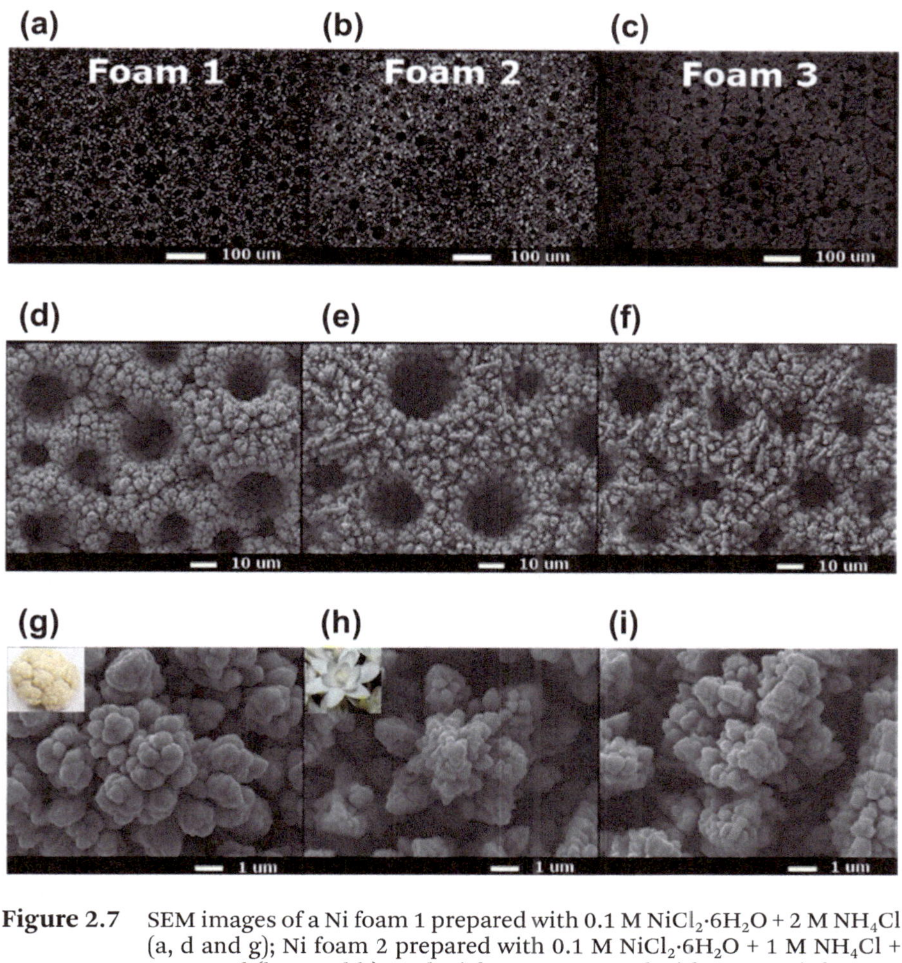

Figure 2.7 SEM images of a Ni foam 1 prepared with 0.1 M $NiCl_2{\cdot}6H_2O$ + 2 M NH_4Cl (a, d and g); Ni foam 2 prepared with 0.1 M $NiCl_2{\cdot}6H_2O$ + 1 M NH_4Cl + 1 M NaCl (b, e and h); and Ni foam 3 prepared with 0.1 M $NiCl_2{\cdot}6H_2O$ + 1 M NH_4Cl (c, f and i). The insets on g and h show a cauliflower and a flower, respectively. Reproduced from ref. 46 with permission from Elsevier, Copyright 2019.

surface porosity and lower pore-size distribution, with a change in the morphological features (from dendritic to cauliflower-like).

The DHBT method is an extremely versatile technique to produce porous structures, and, due to its simplicity and template-free strategy, it has been intensively studied. It is also a simple route to create multi-metal foams as reported in different works – see Table 2.2. Despite these advantages, the DHBT method has some drawbacks concerning the control and extension of the HER reaction and the type of metals that can be deposited on a substrate. Although several metals – including transition metals such as Sn, Fe, Co, Cu, Ni, Mn – have been used for producing metal foams under the cathodic regime, there are still many metals that cannot be deposited because the HER reaction occurs in the same potential and current densities range as the metal.[66] Also, by working at the cathodic potential in the hydrogen evolution region, many parasitic processes highly dependent on several unforeseen parameters are likely to occur. Many of these are now being studied and the underlying mechanisms of foam formation constitute an extensive and intriguing research area.

To create an effective foam the bare substrate must be highly conductive to ensure good adhesion and effective electron flux during the electrodeposition process. Self-standing porous structures have not been produced yet, at least at large scale, limiting the industrial applicability of these materials. Nevertheless, the versality of the process makes it extremely attractive for the production of foam-based electrodes for hydrogen production, for electrocatalytic surfaces, and for electrochemical energy storage, specifically pseudocapacitors and batteries.

2.2 Combustion Synthesis

Combustion synthesis (CS), also known as self-propagating high-temperature synthesis (SHS), is a technique that uses an exothermic reaction that acts as a starting point for a subsequent self-sustaining reaction (*e.g.*, chemical, electrical, mechanical or laser ignition).[67–69] This method has been used to prepare several materials including power ceramics, composites, foams and functionalized materials, due to its high efficiency, inexpensive and low energy consumption.[67,70] In order to induce the combustion reaction, the SHS process requires the presence of an oxidizer, a fuel (reducer) and temperature. The generated heat from a strong exothermic reaction between an organic molecule and a metal complex increases the temperature of the reactive molecules and, subsequently, the neighbouring zone, triggering the propagation of the heat energy as a combustion wave (Figure 2.8).[71] The rapid temperature increase is essential for the development of the crystallinity and purity of the sample. Various classifications of SHS have been proposed by several review authors,[67–69,72–77] depending on the nature of the reactants and the adiabatic temperature: solution combustion synthesis, sol–gel combustion, gas-phase (flame) synthesis and volume combustion. Among these, solution combustion synthesis

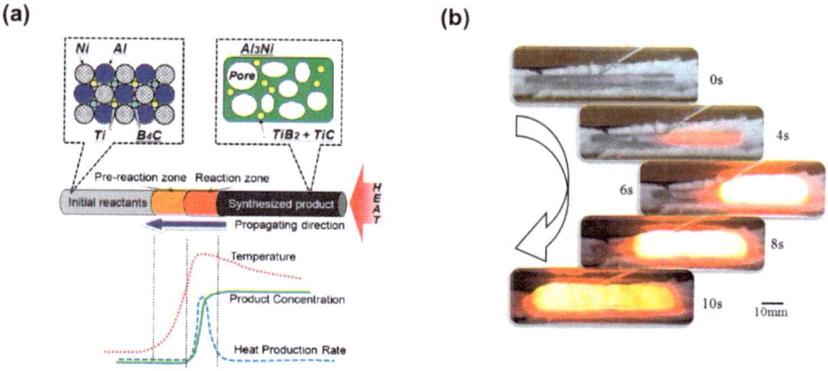

Figure 2.8 Self-propagating combustion synthesis: (a) synthesis of porous Al–Ni intermetallics by self-propagating high-temperature synthesis foaming process, highlighting the reaction wave through the heterogeneous mixture of reactants; (b) self-propagation of the foaming process driven by an exothermic reaction. Reproduced from ref. 78, https://doi.org/10.3390/ma2042360, under the terms of the CC BY 3.0 license, http://creativecommons.org/licenses/by/3.0/.

(SCS) is an important technique due to its versatile experimental parameters on the formation of tailored metal foams with high purity, crystallinity and homogeneity, in a short period of time.[77] The porous structure of metal foams is then formed as the gases resultant from the combustion reaction are generated, favouring the formation of high specific surface area materials.[77,78]

The SCS method consists of a self-sustained exothermic redox reaction on aqueous or sol–gel solutions with a small volume (~1 mm^3).[72] Being a modification of SCS, sol–gel combustion methodology combines the gel transition and combustion processes and yields the formation of metal oxides, alloys and metal foams.[67] It usually employs metal nitrates as precursor oxidizers (lower decomposition temperature compared with chloride and acetate precursors) and organic molecules as reducers (urea, glycine, glycerol, *etc.*), leading to a reaction that is initiated by applying energy, evaporating water, with a gel formation that is subsequently burned during combustion. An important aspect of this method is the ignition temperature used to trigger the combustion reactions that leads to a strong exothermic reaction. A high combustion temperature decreases the time for the initiation of the reaction and results in more crystalline foams and improved particle size.[76] Usually, the ignition temperature is relatively moderate, above the decomposition temperature of the organic compound varying between 300–500 °C. Other ignition methods have also been reported in the literature, such as flame[71,79] and laser ignition.[80,81]

Generally, the SCS process is described by the following equation:

$$M + F + O_2 \leftrightarrow P + G \tag{2.10}$$

where M is the metallic substrate or oxidizer (*e.g.*, metal nitrite), F is the organic fuel or reducer (urea, glycine, glycerol, *etc.*), P is the final metal product and G designates the gases (typically CO_2, H_2O and N_2) produced during the combustion of the fuel that will act as a blowing or gasifying agent.[81] It must be noted that the reaction should be in a controlled atmosphere for the combustion reaction to take place; however, more in-depth investigation is still required to fully understand its influence.[73] The SHS method has been used to prepare transition-metal porous foams.[70,71,78–80,82–89] Several experimental conditions directly affect the micro- and nanostructure of the produced foam, particularly the nature of the precursor and the fuel/oxidizer ratio. For instance, Erri *et al.*[86] synthesized a nickel foam by combusting glycine as fuel and nickel nitrite as oxidizer in an air atmosphere, as described by the stoichiometric equilibrium reaction presented in eqn (2.11):

$$\begin{aligned}\mathrm{Ni(NO_3)_4} + 1.11\phi\ \text{glycine} + 2.50(\phi-1)\,\mathrm{O_2} \leftrightarrow \mathrm{NiO\,(s)} + 2.22\phi\ \mathrm{CO_2} \\ + 2.78\phi\ \mathrm{H_2O} + [1+0.56\phi]\ \mathrm{N_2}\end{aligned} \tag{2.11}$$

where ϕ describes the fuel/oxidizer ratio that controls the propagation of the flame and the overall combustion process. The ratio $\phi = 1$ implies that the atmospheric oxygen is not required for a complete oxidation of the fuel, whereas $\phi > 1$ and $\phi < 1$ conditions indicate an environment fuel-rich system and a fuel-deficient mixture, respectively.[73,75] For $\phi > 1$, the velocity of the combustion reaction progress is low with a slow burning behaviour of the sample and low temperature reaction. This gives extra control on the wave propagation and, consequently, on the structural properties of the foam (Figure 2.9).[82,86] Thus, a variation of ϕ implies a modification of the amount of heat released during the reaction, and consequently a variation of the crystallinity and purity of the final product. This was observed for the Ni foam that revealed a 99.2% of metal at $\phi = 1.3$, with a surface area of the material, in the range of 0.8 to 3.5 $m^2\ g^{-1}$.[86] Moreover, the higher value of fuel/oxidizer ratio induced the production of larger amounts of gases that generally prevents agglomeration.

An important work on the combustion synthesis of metal foams was described by Tappan *et al.*[82] with the development of a combustion synthesis involving metal complexes with nitrogen-containing ligands, particularly bis[1(2)*H*-tetrazol-5-yl]amine or bistetrazolamine (BTA). The decomposition of the complex metal bistetrazolamine (M-BTA) is ignited under an inert atmosphere, resulting in the formation of nanofoams with pores ranging from 5 to 50 nm, as shown in Figure 2.10(a).[90] Several metal ions were synthesized to form monolithic nanoporous metal foams using M-BTA complexes, namely, M = Fe, Co, Cu and Ag. Figure 2.10(b) depicts the ignition of ammonium tris(bis(tetrazolato)amine)ferrate(III) complex (Fe-BTA), resulting in the formation of a Fe metal foam with pore size of a few nanometres (20–200 nm, BET surface area of 258 $m^2\ g^{-1}$) to several microns (1–3 μm), and a BET surface area between 12 to 17 $m^2\ g^{-1}$.[82] In this case, the nitrogen-ligand worked as both blowing and heat agent on a molecular level during the

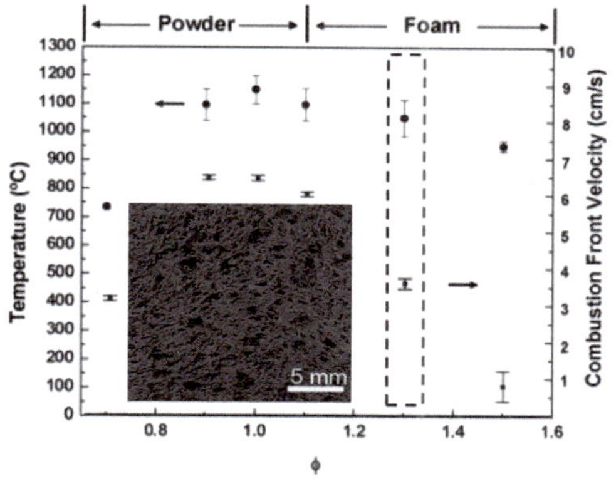

Figure 2.9 Representation of the dependence of combustion velocity and temperature on the fuel-oxidizer ratio ϕ. Inset image: SEM image of the nickel foam synthesized with $\phi = 1.3$. Reproduced from ref. 86 with permission from John Wiley & Sons, Copyright 2008.

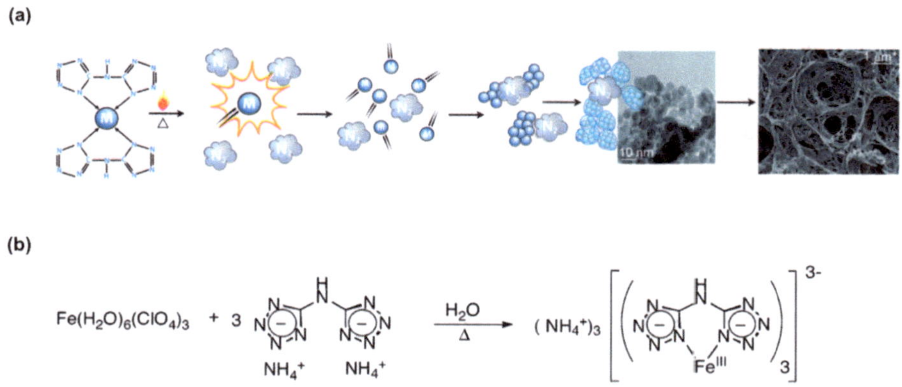

Figure 2.10 Combustion synthesis of metal complexes containing a high-nitrogen energetic ligand. (a) Illustration of the dynamic assembly of nanoporous metal foams by combustion synthesis of metal bistetrazolamine (BTA) complexes. Reproduced from ref. 90 with permission from John Wiley and Sons, Copyright 2010. (b) Complexation reaction to form ammonium tris(bi(tetrazolato)amine)ferrate(III) complex (FeBTA). Reproduced from ref. 82 with permission from American Chemical Society, Copyright 2006.

Fe-BTA decomposition, forming a 3D network. Similar results were observed with the other transition metals.[82]

Other methodologies based on combustion synthesis have also been reported. For example, Kobashi *et al.*[78,84,85] reported the use of titanium and boron carbide (B_4C) to activate Al-TiH_2. The energy generated by this reaction propagated spontaneously through the sample, melting Al[85] and Al–Ni[78,84] nanoparticles and releasing hydrogen gas from the decomposition of TiH_2, producing the pore structure. Moreover, fluorinated organic ligands, such as perfluoroalkyl carboxylic acid ($C_{13}F_{27}COOH$) or polytetrafluoroethylene (Teflon, $(C_2F_4)_n$) have been used as gasifying agents to produce highly porous metallic Al–Ti and Al–Ni foams.[80,88] The insertion of gas-generating additives exhibited different combustion velocities that induced various degrees of porosity. More recently, Karczewski *et al.*[89] reported the synthesis of Fe–Al foams by elemental iron and aluminium powders sintering with tartaric acid additive that worked as an *in situ* gas-releasing agent during SHS. The Fe–Al intermetallic foam showed regular distribution of the pores, with an average diameter of 35 to 47 µm.

Despite the advantages of SHS, combustion synthesis routes exhibit some drawbacks, particularly powder agglomeration, poor control of morphology of the obtained foam and presence of organic impurities associated with an incomplete combustion of the fuel.[76] Thus, there is still the need to further investigate and optimize this process to increase both the efficiency and quality of the obtained foams.

2.3 Industrial Perspectives on the Production of Metal Foams

From an engineering and industrial point of view, the methods to produce foams should be scaled in a reliable and cost-effective way. However, the major problem of the industrial production of foams has been reported to be the lack of economically-viable solutions.[91] Nonetheless, several commercial production methods have been used to produce metal foams, targeting specific applications of the foams at a reasonable price. Closed-cell metal foams are usually obtained by the liquid metallurgy route (ML – melt and PM – powder metallurgical route) due to its industrial simplicity and low cost.[91,92] The ML method consists of blowing gas into molten metal or by adding a blowing agent to the molten metal that releases gas upon decomposition, forming the foam structure. Although economically good, this metallurgic route has some major disadvantages, particularly, the poor control of the gas dispersion, dependency on energy intensive processes, non-uniform pore size and pore distribution in the foam.[92] In contrast to ML, PM gives better control over the process parameters and, hence, better structural, mechanical and physical features to the resulting foam. One example of PM is the synthesis of Al- and Ni–Al intermetallic foams reported by Kobashi *et al.*[78,84,85] that was discussed in the previous section. Open-cell foams have been commercially prepared by a modified templating technique, in which the final morphology of the metal foam is a replica of a scaffold, which can be a polymer, *e.g.*, polyurethane foam. This polymeric template is baked in order to form a negative image of the foam by decomposition, and thereafter filled with molten metal to produce the open-cell foam (Figure 2.11).[93] Nowadays, this alternative casting method (also known as lost-wax casting) has been used

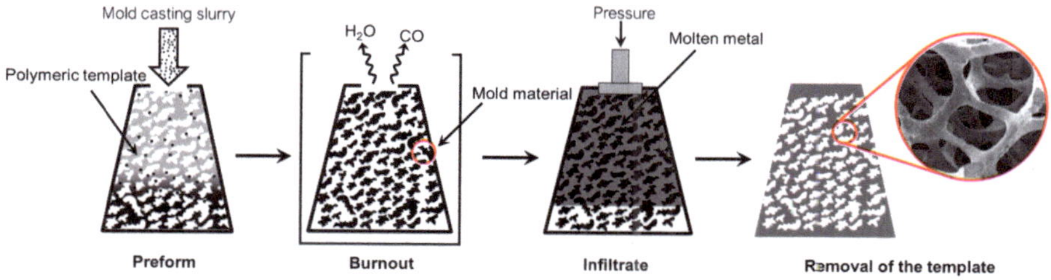

Figure 2.11 Schematic illustration of the casting method.

on the synthesis of metal foams following an analogous method but using wax instead. Despite the formation of highly porous foams, it has several drawbacks concerning the number of production steps, increased price of production and the dimensions of the pore structure, *etc.*[93]

Thus, metal foams also fulfil an important area in modern Li-ion batteries. The main advantage relies on their high surface area, attractive weight/volume features and interesting mechanical properties that can better accommodate the stress generated in the electrode materials. There are, however, some gaps to overcome, namely mechanical stability, efficient surface functionalization and optimal film thickness. Also, the pore structure and wall nanoporosity and structure require additional efforts to maximize the performance of the batteries.

References

1. T. A. Rebbecchi and Y. Chen, *J. Mater. Res.*, 2017, **33**, 2–15.
2. C.-E. Kim, R. M. Rahimi and D. F. Bahr, *MRS Commun.*, 2020, **10**, 286–291.
3. K. Ji, C. Xu, H. Zhao and Z. Dai, *J. Power Sources*, 2014, **248**, 307–316.
4. K. Y. Song, G. S. Jang, J. Tao, J. H. Lee and S. K. Joo, *J. Electrochem. Soc.*, 2016, **163**, A2981–A2987.
5. X. Wu, G. He and Y. Ding, *Electrochem. Energy Rev.*, 2020, **3**, 541–580.
6. X. Guo, J. Han, P. Liu, L. Chen, Y. Ito, Z. Jian, T. Jin, A. Hirata, F. Li, T. Fujita, N. Asao, H. Zhou and M. Chen, *Sci. Rep.*, 2016, **6**, 33466.
7. Z. Dan, F. Qin, Y. Sugawara, I. Muto and N. Hara, *Microporous Mesoporous Mater.*, 2013, **165**, 257–264.
8. C. Rurainsky, A. G. Manjón, F. Hiege, Y.-T. Chen, C. Scheu and K. Tschulik, *J. Mater. Chem. A*, 2020, **8**, 19405–19413.
9. J. Zhang, Y. Zhan, H. Bian, Z. Li, C.-K. Tsang, C. Lee, H. Cheng, S. Shu, Y. Y. Li and J. Lu, *J. Power Sources*, 2014, **257**, 374–379.
10. Z. Qi and J. Weissmüller, *ACS Nano*, 2013, **7**, 5948–5954.
11. A. Y. Chen, J. W. Wang, Y. Wang, Y. Q. Jia, J. F. Gu, X. F. Xie and D. Pan, *Electrochim. Acta*, 2015, **153**, 552–558.
12. F. Ruffino, V. Torrisi, R. Grillo, G. Cacciato, M. Zimbone, G. Piccitto and M. G. Grimaldi, *Superlattices Microstruct.*, 2017, **103**, 28–47.

13. Y. C. Liu, J. C. Huang, X. Wang, M. T. Tsai and Z. K. Wang, *RSC Adv.*, 2018, **8**, 13075–13082.
14. T. Y. Wu, X. Wang, J. C. Huang, W. Y. Tsai, Y. Y. Chu, S. Y. Chen and X. H. Du, *Metall. Mater. Trans. B*, 2015, **46**, 2296–2304.
15. X. Wang, Z. Qi, C. Zhao, W. Wang and Z. Zhang, *J. Phys. Chem. C*, 2009, **113**, 13139–13150.
16. Z. Zhang, Y. Wang, Z. Qi, C. Somsen, X. Wang and C. Zhao, *J. Mater. Chem.*, 2009, **19**, 6042–6050.
17. Z. Zhang, Y. Wang, Z. Qi, J. Lin and X. Bian, *J. Phys. Chem. C*, 2009, **113**, 1308–1314.
18. Q. Zhang, X. Wang, Z. Qi, Y. Wang and Z. Zhang, *Electrochim. Acta*, 2009, **54**, 6190–6198.
19. W. B. Liu, S. C. Zhang, N. Li, J. W. Zheng and Y. L. Xing, *Corros. Sci.*, 2011, **53**, 809–814.
20. R. Mao, S. Liang, X. Wang, Q. Yang and B. Han, *Corros. Sci.*, 2012, **60**, 231–237.
21. Q. Yun, Y. B. He, W. Lv, Y. Zhao, B. Li, F. Kang and Q. H. Yang, *Adv. Mater.*, 2016, **28**, 6932–6939.
22. H. Zhao, D. Lei, Y.-B. He, Y. Yuan, Q. Yun, B. Ni, W. Lv, B. Li, Q.-H. Yang, F. Kang and J. Lu, *Adv. Energy Mater.*, 2018, **8**, 1800266.
23. C. Qin, Y. Zhang, Z. Wang, H. Xiong, H. Yu and W. Zhao, *J. Mater. Sci.: Mater. Electron.*, 2016, **27**, 9206–9215.
24. Y. Sun, Y. Ren and K. Yang, *Mater. Lett.*, 2016, **165**, 1–4.
25. Z. Dan, F. Qin, Y. Sugawara, I. Muto and N. Hara, *Intermetallics*, 2012, **29**, 14–20.
26. P. Jenei, C. Kádár, G. Han, P. T. Hung, H. Choe and J. Gubicza, *Metals*, 2020, **10**, 1128.
27. S. Nam, H. Jo, H. Choe, D. Ahn and H. Choi, *Mater. Trans.*, 2014, **55**, 1414–1418.
28. H. Jo, Y.-H. Cho, M. Choi, J. Cho, J. H. Um, Y.-E. Sung and H. Choe, *Mater. Chem. Phys.*, 2014, **145**, 6–11.
29. S. J. Nam, J. Y. Hwang, H.-K. Kim and H. J. Choi, *Acta Mater.*, 2016, **113**, 170–179.
30. G. Han, J. H. Um, H. Park, K. Hong, W.-S. Yoon and H. Choe, *Scr. Mater.*, 2019, **163**, 9–13.
31. W. B. Liu, S. C. Zhang, N. Li, J. W. Zheng and Y. L. Xing, *J. Electrochem. Soc.*, 2011, **158**, D611–D615.
32. W. Liu, C. Xin, L. Chen, J. Yan, N. Li, S. Shi and S. Zhang, *RSC Adv.*, 2016, **6**, 2662–2670.
33. Z. Qi, C. Zhao, X. Wang, J. Lin, W. Shao, Z. Zhang and X. Bian, *J. Phys. Chem. C*, 2009, **113**, 6694–6698.
34. W. B. Liu, S. C. Zhang, N. Li, J. W. Zheng, S. S. An and Y. L. Xing, *Corros. Sci.*, 2012, **58**, 133–138.
35. A. J. Bard, R. Parsons and J. Jordan, *Standard Potentials in Aqueous Solutions*, M. Dekker, New York, 1985.
36. L. Sun, C.-L. Chien and P. C. Searson, *Chem. Mater.*, 2004, **16**, 3125–3129.

37. C. Zhao, X. Wang, Z. Qi, H. Ji and Z. Zhang, *Corros. Sci.*, 2010, **52**, 3962–3972.
38. G.-F. Yang, K.-Y. Song and S.-K. Joo, *RSC Adv.*, 2015, **5**, 16702–16706.
39. Q. Zhang and Z. Zhang, *Phys. Chem. Chem. Phys.*, 2010, **12**, 1453–1472.
40. M. Mirzaee and C. Dehghanian, *J. Solid State Electrochem.*, 2018, **22**, 3639–3645.
41. S. N. Wu, Y. F. Qi, Q. Wang, X. G. Wang, X. J. Zhao and E. C. Yang, *ChemElectroChem*, 2020, **7**, 684–690.
42. B. J. Plowman, L. A. Jones and S. K. Bhargava, *Chem. Commun.*, 2015, **51**, 4331–4346.
43. H. Yang, X. Hao, J. Tang, W. Jin, C. Liu, H. Hou, X. Ji and J. Hu, *Appl. Surf. Sci.*, 2019, **494**, 731–739.
44. D. Nam, R. Kim, D. Han, J. Kim and H. Kwon, *Electrochim. Acta*, 2011, **56**, 9397–9405.
45. J. Niu, X. Liu, K. Xia, L. Xu, Y. Xu, X. Fang and W. Lu, *Int. J. Electrochem. Sci.*, 2015, **10**, 7331–7340.
46. K. I. Siwek, S. Eugénio, D. M. F. Santos, M. T. Silva and M. F. Montemor, *Int. J. Hydrogen Energy*, 2019, **44**, 1701–1709.
47. K. I. Siwek, S. Eugénio, T. M. Silva and M. Fatima Montemor, *Energy Technol.*, 2019, **7**, 1801139.
48. G. A. Lange, S. Eugénio, R. G. Duarte, T. M. Silva, M. J. Carmezim and M. F. Montemor, *J. Electroanal. Chem.*, 2015, **737**, 85–92.
49. K. I. Siwek, S. Eugénio, T. M. Silva and M. F. Montemor, *Electrochim. Acta*, 2018, **281**, 39–47.
50. N. Taherian and S. Sanjabi, *Surf. Rev. Lett.*, 2019, **27**, 1950122.
51. I. Aldama, K. I. Siwek, J. M. Amarilla, J. M. Rojo, S. Eugénio, T. M. Silva and M. F. Montemor, *Journal Energy Storage*, 2019, **22**, 345–353.
52. H. Singh, P. B. Dheeraj, Y. P. Singh, G. Rathore and M. Bhardwaj, *J. Electroanal. Chem.*, 2017, **785**, 1–7.
53. K. Zhang, W. Xiao, J. Li, J. Liu and C. Yan, *Electrochim. Acta*, 2017, **228**, 422–427.
54. T. Nguyen, M. Boudard, M. J. Carmezim and M. F. Montemor, *Electrochim. Acta*, 2016, **202**, 166–174.
55. D. S. P. Cardoso, S. Eugénio, T. M. Silva, D. M. F. Santos, C. A. C. Sequeira and M. F. Montemor, *RSC Adv.*, 2015, **5**, 43456–43461.
56. S. Eugénio, D. S. P. Cardoso, D. M. F. Santos, B. Šljukić and M. F. Montemor, *Int. J. Hydrogen Energy*, 2016, **41**, 14370–14376.
57. K. Zhuo, M.-G. Jeong, M. S. Shin, W. W. Chun, J. W. Bae, P. J. Yoo and C.-H. Chung, *Appl. Surf. Sci.*, 2014, **322**, 15–20.
58. P. Arévalo-Cid, A. Adán-Más, T. M. Silva, J. A. Rodrigues, E. Maçôas, M. F. Vaz and M. F. Montemor, *Mater. Charact.*, 2020, **169**, 110598.
59. L. D. Rafailović, C. Gammer, C. Rentenberger, T. Trišović, C. Kleber and H. Peter Karnthaler, *Nano Energy*, 2013, **2**, 523–529.
60. S. Eugénio, T. M. Silva, M. J. Carmezim, R. G. Duarte and M. F. Montemor, *J. Appl. Electrochem.*, 2013, **44**, 455–465.

61. I. Herraiz-Cardona, E. Ortega, L. Vázquez-Gómez and V. Pérez-Herranz, *Int. J. Hydrogen Energy*, 2012, **37**, 2147–2156.
62. W. Zhang, C. Ding, A. Wang and Y. Zeng, *J. Electrochem. Soc.*, 2015, **162**, D365–D370.
63. D. M. F. Santos, S. Eugénio, C. A. C. Sequeira and M. F. Montemor, *ECS Trans.*, 2015, **64**, 1–16.
64. H. Zhang, Y. Ye, R. Shen, C. Ru and Y. Hu, *J. Electrochem. Soc.*, 2013, **160**, D441–D445.
65. P. Arévalo-Cid, M. F. Vaz and M. F. Montemor, *Mater. Charact.*, 2022, **193**, 112311.
66. K. I. Popov, S. S. Djokić, N. D. Nikolić and V. D. Jovic, *Morphology of Electrochemically and Chemically Deposited Metals*, Springer, Switzerland, 2016.
67. A. S. Mukasyan, A. S. Rogachev and S. T. Aruna, *Adv. Powder Technol.*, 2015, **26**, 954–976.
68. K. C. Patil, S. T. Aruna and T. Mimani, *Curr. Opin. Solid State Mater. Sci.*, 2002, **6**, 507–512.
69. F. Deganello and A. K. Tyagi, *Prog. Cryst. Growth Charact. Mater.*, 2018, **64**, 23–61.
70. A. Khort, K. Podbolotov, R. Serrano-García and Y. K. Gun'ko, *J. Solid State Chem.*, 2017, **253**, 270–276.
71. A. Moloodi, R. Raiszadeh, J. Vahdati-Khaki and A. Babakhani, *J. Alloys Compd.*, 2009, **487**, 413–419.
72. A. Varma, A. S. Mukasyan, A. S. Rogachev and K. V. Manukyan, *Chem. Rev.*, 2016, **116**, 14493–14586.
73. E. Carlos, R. Martins, E. Fortunato and R. Branquinho, *Chemistry*, 2020, **26**, 9099–9125.
74. K. C. Patil, S. T. Aruna and S. Ekambaram, *Curr. Opin. Solid State Mater. Sci.*, 1997, **2**, 158–165.
75. O. Thoda, G. Xanthopoulou, G. Vekinis and A. Chroneos, *Adv. Eng. Mater.*, 2018, **20**, 1800047.
76. E. Novitskaya, J. P. Kelly, S. Bhaduri and O. A. Graeve, *Int. Mater. Rev.*, 2020, 1–27.
77. S. Specchia, G. Ercolino, S. Karimi, C. Italiano and A. Vita, *Int. J. Self-Propag. High-Temp. Synth.*, 2017, **26**, 166–186.
78. M. Kobashi and N. Kanetake, *Materials*, 2009, **2**, 2360–2368.
79. D. Ressnig and M. Antonietti, *Chem. Mater.*, 2014, **26**, 4064–4067.
80. C. Farley, T. Turnbull, M. L. Pantoya and E. M. Hunt, *Acta Mater.*, 2011, **59**, 2447–2454.
81. E. M. Hunt, M. L. Pantoya and R. J. Jouet, *Intermetallics*, 2006, **14**, 620–629.
82. B. C. Tappan, M. H. Huynh, M. A. Hiskey, D. E. Chavez, E. P. Luther, J. T. Mang and S. F. Son, *J. Am. Chem. Soc.*, 2006, **128**, 6589–6594.
83. L. J. Groven and J. A. Puszynski, *Mater. Lett.*, 2012, **73**, 126–128.
84. M. Kobashi and N. Kanetake, *Adv. Eng. Mater.*, 2002, **4**, 745–747.
85. M. Kobashi and N. Kanetake, *Mater. Sci. Forum*, 2006, **519–521**, 1335–1340.
86. P. Erri, J. Nader and A. Varma, *Adv. Mater.*, 2008, **20**, 1243–1245.

87. Z. Zhang, P. He, X. Li, Z. Yang, Y. Fu and Z. Lin, *J. Alloys Compd.*, 2018, **758**, 38–44.
88. O. K. Kamynina, S. G. Vadchenko, A. E. Sytschev, A. S. Rogachev, L. M. Umarov and N. V. Sachkova, *Int. J. Self-Propag. High-Temp. Synth.*, 2007, **16**, 137–140.
89. K. Karczewski, W. J. Stepniowski and M. Salerno, *Materials*, 2018, **11**, 621.
90. B. C. Tappan, S. A. Steiner 3rd and E. P. Luther, *Angew. Chem., Int. Ed.*, 2010, **49**, 4544–4565.
91. F. Garcia-Moreno, *Materials*, 2016, **9**, 85.
92. A. Kulshreshtha and S. K. Dhakad, *Mater. Today: Proc.*, 2020, **26**, 1784–1790.
93. M. F. Ashby, T. Evans, N. A. Fleck, J. W. Hutchinson, H. N. G. Wadley and L. J. Gibson, *Metal Foams: A Design Guide*, Elsevier, 2000.

CHAPTER 3

Functionalization of Metal Nanofoams

3.1 Introduction

Nanostructuring is a widely adopted strategy to enhance the performance of electrodes for different applications. For ordinary powdered materials, the preparation of electrodes usually follows a common coating method consisting of applying the active material on top of the current collector substrate by using a polymeric binder that enables preparation of a slurry (*e.g.*, Nafion, polytetrafluoroethylene, polyvinylidene fluoride, *etc*).[1] Such a strategy leads to undesired additional resistances, introduces inert mass into the electrode, and inhibits mass transport in the electrode. Moreover, problems with low adhesion between the active material and the substrate involving different processing steps are also generally observed. For instance, often, the active material is pressed or calendered onto the surface of the substrate, which might lead to de-adhesion and peeling of the material during long-term or large current demands.[2] Interestingly, these materials can be pressed into the porous structure of metallic foams (*e.g.*, Ni foams) but using a slurry hardly takes full advantage of the high surface area and porosity of the metal foams. This happens, for example, in battery assembly when Ni foams are used as a current collector. Hence, the functionalization of metallic foams becomes an attractive strategy to meet the needs of various modern applications.[2–4] In this case, the metal foam is employed as a scaffold material that serves as a support for an active material either obtained from the foam itself (*e.g.*, by oxidation) or by adding a new compound onto the foam substrate.

Transition Metal-based Nanofoams for Electrochemical Systems: Manufacturing, Characterization and Applications
By Maria de Fátima Montemor, Diana M. Fernandes, Alberto Adan-Mas, Ana Catarina Alves, Gabriel Garcia Carvalho and Inês S. Marques

Published by the Royal Society of Chemistry, www.rsc.org

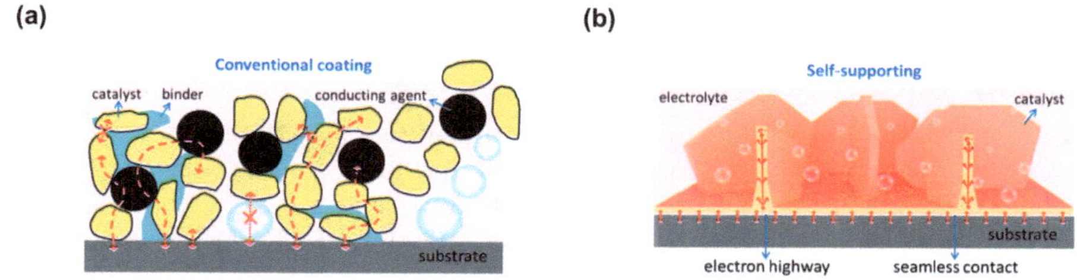

Figure 3.1 Schematic views of the structural and electrochemical features of (a) conventional powdery catalysts coated on current collectors with polymer binder and carbon additives and (b) self-supported electrode with *in situ* grown catalysts on the substrates. Reproduced from ref. 1 with permission from John Wiley and Sons, Copyright 2019.

This can be done in such a way that it is possible to create a thin nanometric film, mimicking the original foam structure.

Commercially available Ni foams have been extensively used as catalytic substrates due to their excellent conductivity, specific and electrochemically accessible surface area, high porosity and low cost.[5–10] Such Ni foam features are of great interest for nickel–metal hydride[11] and Li-ion batteries,[12] for example. The functionalization of Ni foams is generally based on the *in situ* growth of an active material on the foam surface, thereby maximizing the number of reaction sites of the electrodes. In this way, a good mechanical adhesion is achieved, and a binder-free strategy enables a better contact between the electrode substrate and the active material, leading to reduced contact resistance of the system (Figure 3.1). One- or multi-step hydrothermal, solvothermal, chemical bath and electrodeposition procedures have been widely reported for functionalizing commercial porous metal nanofoams.[6–10,13–77] In the following sections, we will briefly discuss the main techniques used for the functionalization of metallic foams used as electrodes in electrochemical energy conversion and storage applications.

3.2 Hydrothermal Methods

Hydrothermal techniques refer to a heterogeneous chemical reaction in aqueous solution at relatively high and controlled conditions of temperature and pressure.[78–80] Since the nineteenth century, geological sciences have been using hydrothermal techniques to synthesize and extract elements from minerals. Currently, the hydrothermal method is one of the most used strategies for surface functionalization by directly growing active materials onto metallic substrates, resulting in a uniform deposition of nanostructures. Therefore, this technology has become essential in the preparation of high-quality materials with tailored size, shape and surface properties, and, in many cases, in an environmentally-friendly solvent.

Hydrothermal synthesis is based on the precipitation (or nucleation and growth) of metal nanostructures on the metal foam surface at high temperature and pressure conditions. During this process, the pre-synthesized material can undergo chemical and morphological changes induced by conversion chemical reactions, including anion exchange reactions and the Kirkendall effect (exemplified in Figure 3.2), leading to the formation of oxide, phosphide, sulfide and selenide composites.[81] The anion exchange reaction consists of the replacement of anions in the multicomponent template. Associated with this process, the Kirkendall effect might be observed, leading to the obtention of porous structures. The Kirkendall effect is controlled by the diffusion rate of species. In fact, the different intrinsic diffusivities of the chemical components have been used to guide the growth of porous structures by means of this technique.[37,46]

Extensive research has been reported concerning the functionalization of 3D nanoporous metal foams using hydrothermal techniques in the last decade.[6–8,10,13–54] Moreover, several approaches have been described, based on different experimental conditions, particularly varying the methodology used (one *vs.* multi-step reaction), time and temperature of reaction and the addition of inorganic/organic additives to the solution (ammonium fluoride, sodium sulfide, *etc*).

An important aspect of the hydrothermal method is its versatility, giving rise to one- or multi-step strategy reactions. One-step synthesis routes of metal foams have attracted much attention due to their simplicity, low cost and

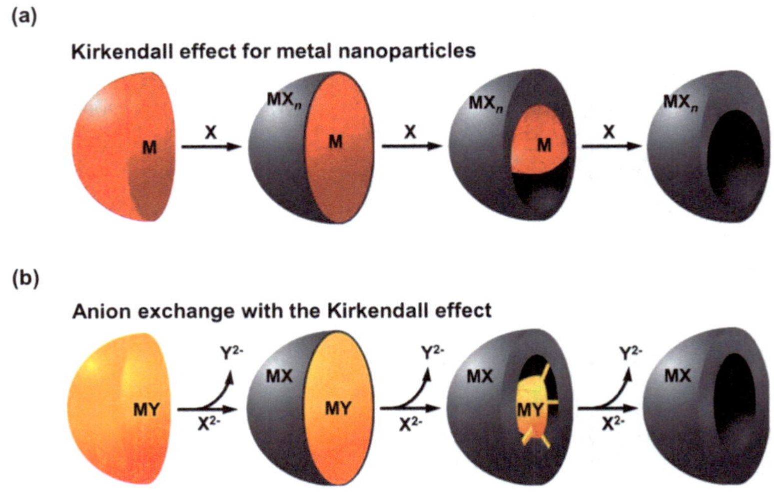

Figure 3.2 Examples of nanoparticles' conversion chemistry: (a) Kirkendall effect: during the reaction of the metal M with reactive species X to form MX_n, void formation occurs. (b) Anion exchange: during the reaction of a metal compound MY with a reactive anion, X^{2-} are released to form MX; void formation through the Kirkendall effect can occur when cations diffuse outward to react with the reactive anions. Reproduced from ref. 81 with permission from the Royal Society of Chemistry.

speed of preparation of the nanostructured materials with fine growth control. In light of these considerations, facile one-step hydrothermal synthesis procedures have been proposed in the literature.[6,13,19,23,24,35,36,38,42,45,47,50,53,82] Among others, nanostructured transition metal sulfides (NiS_x, CoS_x, MoS_x, *etc.*) have been intensively studied in many energy-related applications, from supercapacitor and Li-ion batteries to hydrogen production and electrochemical sensors.[13,19,24,36,38,50,53] The experimental parameters reported in these works are summarized in Table 3.1. There are also multi-step routes that combine two or more procedures to functionalize metal foams, for example, hydrothermal routes followed by phosphorization,[7,22,32] electrodeposition,[10,15,27,33] calcination or annealing process,[14,16–18,25,28,29,31,40,49] selenization,[21] sulfurization,[34] or even another hydrothermal step.[8,20,26,30,37,39,41,43,44,46,48,51,52] These approaches have been used to prepare, for instance, core–shell structures by depositing an active shell on a previously grown material on a commercial foam. Commercial foam or a core material is used as a porous backbone to support the active material to ensure effective electrical connection and mechanical strength of the electrode. Table 3.2 summarizes the experimental parameters of the multi-step strategy to functionalize metal foams. Although very versatile, multi-step processes are challenging considering that the original foam materials can be degraded, their properties changed or irreversibly affected. Chemical stability of the original foam is still a major obstacle and important research efforts are still ongoing to use this flexible route to create new surface compositions onto metal foams.

3.3 Other Processes

The functionalization of commercially available metal foams has been investigated by other methods: electrodeposition,[55–62,83–87] thermal methods[9,63–66,68,88] and chemical bath functionalization[67,69–77,89] as summarized in Tables 3.3–3.5.

Electrodeposition is a technique that enables the formation of a thin film directly on a highly porous substrate, conserving the structure of the foam. Electrodeposition is dependent on the deposition potential (or current), composition of the electrolyte solution, and time and temperature of deposition. Usually, the functionalization of metal foams with metal oxides by electrodeposition of metal ions follows a two-step route: firstly, OH^- ions react with metal ions on the electrolyte solution to produce a precipitate of hydroxide/oxyhydroxide by electrodeposition;[59,60] and secondly, the as-formed (oxy)hydroxides are transformed thermally into the corresponding oxides.[55–58] The morphology of the obtained metal oxides is generally composed by nanosheet arrays due to the ultrathin features of the deposited film of oxide. For instance, Siwek *et al.* in ref. 90 electrodeposited Mn oxides with different masses and thicknesses on the top of 3D metallic and bimetallic foams. Such structures were prepared using an as-deposited Ni foam or Ni–Co foams, followed by an anodic electrodeposition of $MnO_x{\cdot}nH_2O$, as depicted in Figure 3.3. One particular strategy for the functionalization of

Table 3.1 Summary of the one-step hydrothermal routes used for the functionalization of metal foams.[a]

Material	Conditions	Temperature/°C	Time/h	Mass loading/ mg cm^{-2}	Morphology	References
Co,Mn-LDH@ Ni foam	Water	120	6	1.25	Nanoneedle 25 nm diameter, 2.8 μm length	53
$NiCo_2S_4$@Ni_3S_2 on Ni foam	Water	160	12	7.875	Ni_3S_2 nanosheets cover the $NiCo_2S_4$ nanorod; 93.7 m^2 g^{-1}	13
MIL-53(FeNi) on Ni foam	DMF/ethanol/water	125	12	2.63	Nanosheets	35
CoS_x/Ni_3S_2@ Ni foam	Water	120	12	2.83	Cuboid-shaped CoS_x	36
MoS_2/Ni_3S_2 on Ni foam	Water	180	24	Not given	Nanowires with 4 μm length	6
Ni_3Se_2 on Ni foam	Water	160	25	Not given	Cauliflower-shaped with 1.5 μm, and pore size 300 nm	23
$CoMoO_4$	Water	180	6	Not given	Nanoplate arrays with 300–500 nm length and 50–100 nm breadth	42
$MnCo_2O_4$ on Ni foam	Water, NH_4F	100	5	1.1	Microsphere with 5 μm diameter	45
NiSe on Ni foam	Water, $NaBH_4$	140	12	2.8	Nanowires with 20–80 nm diameter	47
Ni_3S_2/AT on Ni foam	Water	160	6	Not given	Nanorod with 1–2 μm length 30.9 m^2 g^{-1}	38
Co_9S_8/rGO/Ni_3S_2 on Ni foam	Water	180	24	Not given	Nanoflakes	50
Ni_3S_2 on Ni foam	Water	120	16	2.0	Nanosheets	19
Ni_3S_2 on Ni foam	Water, TAA	180	4	37	Nanorods with 300–600 nm diameter and 1 μm length	24
Ni_3S_2 on Ni foam	Water, thiourea	180	1, 2, 3	Not given	Porous network with pore diameters ranging from 50 to 600 nm	82

[a]LDH: layered double hydroxide; DMF: *N*,*N*-dimethylformamide; TAA: thioacetamide; AT: acid-treated; rGO: reduced graphene oxide.

metal foams using electrodeposition was described by Chen *et al.*,[62] in which the metal foam used acted as a substrate to deposit active materials, but also as a metal source for *in situ* functionalization. In this method, the commercial foam was oxidized at the applied potentials and the metal ions from the foam were chelated with metal phytate groups, to form the final material in a one-step route.[62] Two of the main challenges when functionalizing metal foams by electrodeposition are (i) maintaining the chemical stability of the bare foam in the new electrodeposition conditions and (ii) controlling the mass and thickness of the deposited material. If not well finely tuned, the advantages of using the foam as a scaffold to create a highly porous foam-metal compound 3D architecture can be lost.

Taking advantage of chemical transformations under extreme temperature conditions, thermal routes have also been used to prepare 3D porous materials by *in situ* growth of active material on the backbones of different foams.[9,63–66,68] In this procedure, the atmosphere composition, temperature of the furnace and time of reaction are crucial for tailoring the nanostructure. For example, Zhou *et al.*[9] described the preparation of $NiSe_2$-based materials by partial thermal selenization of a Ni foam, with Se powder as the selenium source. Figure 3.4 shows SEM images of the $NiSe_2$/Ni foam materials prepared by Zhou's group at different temperatures. It can be observed that, with an increasing applied temperature, the surface of the material became rougher until, eventually, it was broken by the $NiSe_2$ expansion on the foam's surface. This change on the roughness of the surface was correlated to the increase of the number of exposed active sites during HER. Thermal phosphorization[63,68] enables the conversion of metal ions into the corresponding metal phosphide, in the presence of a phosphorus source such as sodium hypophosphite monohydrate ($NaH_2PO_2 \cdot H_2O$), while sulfurization uses a sulfur-containing precursor, as reported by other authors to functionalize the foam.[64,66] Thermal routes also carry important challenges. Depending on the conditions, temperature can induce stress to the material, altering the original properties of the bare foam. To optimize the functionalization process, important efforts must be undertaken to better correlate the parameters of the thermal treatments with the final properties of the resulting material.

Chemical bath deposition is a simple process of functionalizing foams by immersion or reflux, soaking the metal substrate in a solution containing precursor species that allow growth of the active material. Usually, wet methods are used due to their simplicity, mild-conditions' synthesis, low cost and template-free route. This method has been used to synthesize several active materials grown on metallic foams, including layered double hydroxides (LDH),[69,76] nitride,[70] hydroxide,[71,74,77] oxide[72,73,75] and bimetallic alloy[67] structures. A general chemical bath deposition is schematically illustrated in Figure 3.5, in which the commercial foam follows a dip-coating step in a solution containing metal ions that are deposited on top of the foam by, for example, precipitation of the corresponding hydroxides.[74] The amount of material grown on the metal foam is controlled by changing the concentration of the metal ion, the pH and temperature of the solution, and the

Table 3.2 Summary of the multi-step hydrothermal routes used for the functionalization of metal foams.[a]

		Hydro- or solvothermal conditions		Chemical bath conditions		
Material	Methodology	Temperature/°C	Time/h	Solution	Temperature/°C	Time/h
P–Co–Ni–S on Ni foams	Three-step hydrothermal	150; 150; 150	11; 11; 11	—	—	—
NiCo-LDH@ Co_9S_8@ Ni foams	Two-step hydrothermal	120; 90	6; 10	—	—	—
MOF $CoSeO_3$ on Ni foam	Hydrothermal; chemical bath; thermal treatment; 2nd hydrothermal	120; 200	5; 24	2-MeI in ethanol and water	RT	12
Co-NCNTFs on Ni foam	Solvothermal (DMF); thermal treatment	115	24	—	—	—
Zn–Ni–Se/ $Ni(OH)_2$ on Ni foam	Chemical bath; two-step hydrothermal	140; 150	10; 6	Water and 2-MeI	RT	4
Ni_2P on Fe foam	Hydrothermal; thermal phosphorization	100	5	—	—	—
$NiGa_2S_4$-rGO on Ni foam	Hydrothermal; electrodeposition of rGO	120	12	—	—	—
NiMn LDH@ Co_3O_4@ Ni foam	Hydrothermal; calcination; chemical bath	120	9	Water	80	2.5
NiFeMoS on Ni foam	Electrodeposition; hydrothermal	200	24	—	—	—
E-$CoSe_2$/ $Ni_{0.85}Se$@ Ni foam	Two-step hydrothermal process; electrochemical activation	120; 120	8; 8	—	—	—
$MgCo_2O_4$@PPy on Ni foam	Hydrothermal; calcination; chemical oxidative polymerization of PPy	120	6	—	—	—
$MnMoO_4{\cdot}H_2O$@ MnO_2 on Ni foam	Two-step hydrothermal	150; 140	8; 3	—	—	—
$Ni_{0.89}Co_{0.11}Se_2$ on Ni foam	Hydrothermal; hydrothermal selenization; acid etching	180; 180	20; 16	—	—	—
$NiCo_2S_4$@NiFe-LDH on Ni foam	Three-step hydrothermal	120; 180; 120	10; 6; 10	—	—	—

Thermal treatment conditions			Electrodeposition conditions			Mass loading/mg cm^{-2}	Morphology	References
Temperature/°C	Time/h	Atmosphere	Solution	Potential	Time			
—	—	—	—	—	—	Not given	Starfish structure	91
—	—	—	—	—	—	1.8	Nanowires	92
350	1	N_2	—	—	—	Not given	Nanowires	93
600	3	Vacuum	—	—	—	Not given	Microsheet arrays	94
—	—	—	—	—	—	5.3	Uniform and compact hexagonal nanosheets on NF surface	54
310	1	N_2	—	—	—	4.96	Nanosheets with 20 nm thickness	32
—	—	—	—	—	—	2.3	Nanoparticles; pore size distribution 7.5 nm; 55.2 $m^2\ g^{-1}$	33
400	4	Air	—	—	—	2–3	Core–shell structure	14
—	—	—	10 mM Ni and Fe salts solution	Potentiostatic at −1 V *vs.* Ag/AgCl	600 s	Not given	Anemones-like nanorods	34
—	—	—	6 M KOH	Potentiostatic at 0.7 V *vs.* Hg/HgO	15 min	5.25	Nanosheets fully covering the surface of nanowires	15
350	2	Air	—	—	—	Not given	Nanowires of 40 nm thickness	16
—	—	—	—	—	—	0.85 to 1.4	Core–shell nanosheets with 40 to 200 nm thickness	20
—	—	—	—	—	—	2.62	Nanosheets with 12 nm thickness; 37.32 $m^2\ g^{-1}$	21
—	—	—	—	—	—	Not given	Nanotubes with 200 nm diameter and wall thickness of 25 nm	39

(*continued*)

Table 3.2 (*Continued*)

Material	Methodology	Hydro- or solvothermal conditions		Chemical bath conditions		
		Temperature/°C	Time/h	Solution	Temperature/°C	Time/h
MoP/Ni_2P on Ni foam	Hydrothermal; thermal phosphorization	100	18	—	—	—
NiCoP on Ni foam	Hydrothermal; thermal phosphorization	120	6	—	—	—
$NiCo_2O_4$@rGO on Ni foam	Hydrothermal; annealing	90	12	—	—	—
$NiCo_2S_4$ on Ni foam	Two-step hydrothermal	95; 120	12; 8	—	—	—
$NiCo_2S_4$ on Ni foam	Two-step hydrothermal	120; 160	8; 8	—	—	—
NiFe/$NiCo_2O_4$ on Ni foam	Hydrothermal; annealing; electrodeposition	120	5	—	—	—
$NiCo_2S_4$@$Co(OH)_2$ on Ni foam	Two-step hydrothermal; electrodeposition	120; 120	6; 14	—	—	—
$NiCo_2S_4$@PPy on Ni foam	Two-step hydrothermal; electrodeposition of PPy	120; 120	6; 14	—	—	—
$CoMoO_4$@$NiMoO_4$ on Ni foam	Two-step hydrothermal; calcination treatment	140; 80	4; 4	—	—	—
$NiCo_2S_4$ on Ni foam	Two-step hydrothermal	100; 110	8; 4	—	—	—
NCO on graphene-nickel foam	Chemical vapour deposition of graphene; hydrothermal; calcination	90	5	—	—	—
$NiCo_2S_4$ on Ni foam	Hydrothermal; annealing; hydrothermal	120; 120	6; 6	—	—	—
$NiCo_2O_4$@PPy on Ni foam	Hydrothermal; calcination; oxidative polymerization of PPy	110	6	—	—	—

Thermal treatment conditions			Electrodeposition conditions					
Temperature/°C	Time/h	Atmosphere	Solution	Potential	Time	Mass loading/mg cm^{-2}	Morphology	References
400	300 min	N_2	—	—	—	Not given	Nanoparticles and nanoflowers distributed on the surface	22
300	3	Ar	—	—	—	8	Cone-shaped nanowire arrays with 80–150 nm diameter	7
300	4	Ar	—	—	—	3.2	Nanosheets with 1.4 μm width	40
—	—	—	—	—	—	4.0	Caterpillar-like with length of 2–5 μm	41
—	—	—	—	—	—	Not given	Nanowire with 80 nm diameter and 1 μm length	30
350	2	Ar	Ni and Fe solution	Potentiostatic at −1 V *vs.* Ag/AgCl	300 s	Not given	Nanoflakes with 500 nm to 2 μm length	10
—	—	—	Co solution	−1.2 V to 0.2 V *vs.* Ag/AgCl, 50 mV s^{-1}	1–3 cycles	9.1	Core–shell nanotube array	43
—	—	—	—	—	—	6.87	Nanotubes	8
350	2	Air	—	—	—	Not given	Nanosheets with 100 nm thickness; 31.77 m^2 g^{-1}	44
—	—	—	—	—	—	1.0	Rod-like nanomaterial with 10 nm thickness; 26.6 m^2 g^{-1}	48
250	3	Air	—	—	—	1.1 to 1.2	Nanoneedles with 300 nm length and 20 nm diameter; 150.314 m^2 g^{-1}	49
400	3	Air	—	—	—	7.5	Nanotubes with 7 μm length	37
300	2	Air	—	—	—	1.86	Nanowires with 80–100 nm diameter and 1–2 μm length	25

(continued)

Table 3.2 (*Continued*)

		Hydro- or solvothermal conditions		Chemical bath conditions		
Material	Methodology	Temperature/°C	Time/h	Solution	Temperature/°C	Time/h
$NiCo_2S_4$ on Ni foam	Two-step hydrothermal	120; 160	6; 6	—	—	—
$CoNi_2S_4$ on Ni foam	Two-step hydrothermal	95; 120	8; 8	—	—	—
$NiCo_2O_4$@MnO_2 on Ni foam	Hydrothermal; annealing; electrodeposition; annealing	120	6	—	—	—
Co_3O_4@MnO_2 on Ni foam	Hydrothermal; annealing; hydrothermal	105; 160	5; 24	—	—	—
$NiCo_2S_4$ on Ni foam	Two-step hydrothermal	100; 120	8; 4	—	—	—
$NiCo_2O_4$@MnO_2 on Ni foam	Two-step hydrothermal; calcination	160	0.5	—	—	—
Ni foam-supported $NiCo_2O_4$	Hydrothermal; calcination	120	6	—	—	—
$NiMoO_4$ on Ni foam	Hydrothermal; calcination	150	6	—	—	—
Co_3O_4 on Ni foam	Hydrothermal; annealing	95	8	—	—	—
CoO@NiHON on Ni foam	Hydrothermal; annealing; chemical bath	95	8	0.1 M $Ni(NO_3)_2$	85 °C	5
$Co(OH)_2$/$CuCo_2S_4$/Ni foam	Two-step hydrothermal; electrodeposition	120; 160	6; 8	—	—	—

[a]LDH: layered double hydroxide; NCNTFs: N-doped carbon nanotube frameworks; VACNTF: vertically-aligned carbon nanotube array film; TAA: thioacetamide; 2-MeI: 2-methylimidazole; DMF: *N*,*N*-dimethylformamide; PPy: polypyrrole.

Thermal treatment conditions			Electrodeposition conditions					
Temperature/°C	Time/h	Atmosphere	Solution	Potential	Time	Mass loading/mg cm^{-2}	Morphology	References
—	—	—	—	—	—	49.5	Nanotubes	26
—	—	—	—	—	—	2.2	Nanosheets with 45 nm thickness; 22.9 m^2 g^{-1}	51
300; 200	2; 2	Not given	0.1 $MnAc_2$ + NH_4Ac	0.5 mA cm^{-2}	4–20 min	Not given	Nanowires with 2–5 μm length and 20–80 nm diameter; 103.491 m^2 g^{-1}	27
350	2	Air	—	—	—	Not given	Co_3O_4 nanowires; MnO_2 nanosheets	17
—	—	—	—	—	—	4.2	Nanorod with 120 nm diameter and 30 nm wall thickness	46
350	2	Not given	—	—	—	1.4	Nanowires	52
400	3	Air	—	—	—	3	Nanowires with 3 μm length and 80–200 nm diameter	28
400	1	Ar	—	—	—	Not given	Nanowires with 4 μm length and 50 nm diameter	29
250	1	Air	—	—	—	2.9	Nanowires with 80 nm diameter and 10 μm length; 74.9 m^2 g^{-1}	31
450	2	Ar	—	—	—	CoO: 2.2NiHON: 0.8	Nanowires with 60–100 nm diameter and 2 μm length	18
—	—	—	0.05 M $Co(NO_3)_2$	−1.0 V	10 min	1.8 mg	Nanoneedles	95

Table 3.3 Summary of the electrodeposition routes used for the functionalization of metal foams.[a]

Material	Methodology	Solution	Potential	Time	Mass loading/mg cm^{-2}	Morphology	References
N7CF-phy on Fe foam	Electrochemical erosion	Co^{2+}, Ni^{2+}, Phy-Na	−2.5 to 1.5 V *vs.* Ag/AgCl	20 mV s^{-1} for 8 cycles	Not given	Granule congeries and wrinkled structures clad inside	62 and 64
NiS_x on Cu foam	Potentiostatic electrodeposition	$NiCl_2{\cdot}H_2O$, CH_4N_2S	−0.65 V *vs.* Ag/AgCl	40 min	Not given	Microsphere-like morphology	61
Ni foam/graphene/$NiCo_2O_4$	Potentiostatic electrodeposition of Ni, Co; calcination (300 °C, 2 h)	Co^{2+}, Ni^{2+}	−1.0 V *vs.* SCE	10 min	Not given	Mesoporous layer of graphene and $NiCo_2O_4$	55
$NiCo_2O_4$ on Ni foam	Co-deposition; calcination (300 °C, 2 h)	Co^{2+}, Ni^{2+}	−1.0 V *vs.* SCE	5 min	0.8	Nanosheet	56
Co_3O_4 on Ni foam	Deposition; calcination (250 °C, 2 h)	Co^{2+}	−1.0 V *vs.* SCE	20 min	1.4	Nanosheets with 10 nm thickness; 118 m^2 g^{-1}	57
NS-CuO on Cu-foam	Anodization; calcination (air atm, 200 °C, 15 min)	—	20 mA cm^{-2}	20 min	12.2	Nanosheets with 20 nm thickness and 340 nm width; 34.8 m^2 g^{-1}	58
$Co(OH)_2$/GNS on Ni foam	Chemical vapour deposition; electrodeposition	Co^{2+}	−0.9 V *vs.* SCE	Not given	0.005	Nanosheets with 12 nm thickness	59
α-$Co(HO)_2$ film on nickel foam	Electrodeposition	Co^{2+}	−1.0 V *vs.* SCE	100 s	Not given	Leaves	60
Ni foam/$MnO_x{\cdot}xH_2O$	Electrodeposition	Mn^{2+}	0 to 5 mA	2500 cycles	1.4 ± 0.05 mg cm^{-2}	Spherical structures	90
			0 to 4 mA	2000 cycles	0.65 ± 0.05 mg cm^{-2}		

MnO_2/graphene/Ni foam	Anodic stripping method; electrodeposition of MnO_2; calcination (250 °C, 2 h, N_2 atm)	Mn^{2+}	1, 5, 10, 15 $mA\ cm^{-2}$	180 s	0.09, 0.12, 0.18, 0.25 mg	Nanoparticles from 10 to 50 nm	83
Porous graphene networks/nickel foam	Electrodeposition	0.15 M $LiClO_4$ + graphene oxide	−1.0, −1.1, −1.2, −1.3 V	300, 400, 500, 600, 700, 800 s	1.0–2.0 mg	Sheets	84
NiCu foams	Electrodeposition	$NiSO_4 \cdot 7H_2O$ (0.5 M), H_2SO_4 (1.5 M), HCl (1 M) and $CuSO_4 \cdot 5H_2O$ (0.01 M)	1.8 $A\ cm^{-2}$	90, 180 s	Not given	Dendritic structure with nanosheet	85
MnO_x@Ni_{comm}	Electrodeposition	$NiCl_2 \cdot 6H_2O$, NH_4Cl, NaCl; $KMnO_4$, $NaSO_4$	−0.39 $A\ cm^{-2}$; −0.5	180 s; 1320–5000 s	10.5 $mg\ cm^{-2}$	3D foam-like architecture	86
MnO_x@Ni_f		$CoCl_2 \cdot 6H_2O$, NH_4Cl, NaCl; $KMnO_4$, $NaSO_4$	to −3 $mA\ cm^{-2}$		10.9 $mg\ cm^{-2}$		
MnO_x@$NiCo_f$		$NiCl_2 \cdot 6H_2O$, $CoCl_2 \cdot 6H_2O$, NH_4Cl, NaCl, $KMnO_4$, $NaSO_4$			10.2 $mg\ cm^{-2}$	Cauliflower-like to fine-grained flower-type morphology	
Cu foam	Electrodeposition	$CuSO_4$, H_2SO_4, $(NH_4)_2SO_4$, BTA	−3 $mA\ cm^{-2}$	4 s	Not given	Sphere-shaped Cu nanoparticles ~5 nm	87
Sn/Cu foam		$SnSO_4$, H_2SO_4, Triton X-100	−15 $mA\ cm^{-2}$	300 s		Uniformly deposited on the porous Cu foam	

[a]GNS: graphene nanosheets; N7C7: chelated with bi-metallic phytates; GNS: graphene nanosheets.

Table 3.4 Summary of the thermal routes used for the functionalization of metal foams.[a]

Material	Method	Atmosphere	Temperature/°C	Time/h	Mass loading/ mg cm^{-2}	Morphology	References
(Fe–Co–Ni) P@PC on Ni foam	Calcination	NaH_2PO_2, N_2 atm	400	2	Not given	Nanosheets	68
Ni–Cr@ Cu foam	Annealing	H_2/Ar atm	500	4	Not given	Ni–Cr hierarchical structures	67
Ni_3S_2 foam Co_9S_8 foam FeS foam	Annealing reaction with thiourea (sulfation); CNTs grown by thermal at 900 °C	Thiourea	600	1	Not given	Metal foamy structure; CNTs with length of ~15–20 μm	66
NCNT on Ni foam	Pre-coating w/ PDA; pyrolysis	Ar atm + 5% H_2	800	2	Not given	Pristine 3D porous structure with NCNTs of 20–80 nm diameter	65
$Fe(PO_3)_2/Ni_2P$ on Ni foam	Phosphidation	Hypophosphite monohydrate under Ar atm	450	1	8	Not given	63
NiS on Ni foam	Sulfurization	Se power under Ar atm	350	Not given	43	Microspheres with a diameter size 0.6–2.0 μm	64
$NiSe_2$ on Ni foam	Thermal selenization	Se power under Ar atm	600	1	Not given	Pore with 100 μm	9
SnS_x@Ni foam	Pyrolysis	Ar atom 99.999%, H_2S reactant gas	180; 160	Different cycles	Not given	Flake/granules-like morphology	88

[a]PC: polydopamine-derived carbon; NCNT: N-doped carbon nanotubes.

Table 3.5 Summary of the chemical routes used for the functionalization of metal foams.[a]

Material	Methodology	Solution	Temperature/°C	Time/h	Mass loading/mg cm^{-2}	Morphology	References
NiF/Fe^{3+} CoF/Fe^{3+}	Hydrolytic reaction	Aqueous solution	RT	1	68	Well-oriented nanosheets	77
pc-NiFe on Ni foam	Two-step growth solution: dip-coating	NaOH aqueous solution	RT	3 and 5 s; dried for 10 and 5 min	0.5	Nanosheets with range 150–600 nm and thickness 10 nm	76
NiMn on Ni foam	Wet method	Water	80	2	0.15–1.2	Nanosheets with 18 to 26 nm thickness	69
$FeNi_3N$ on Ni foam	Chemical precipitation; thermal ammonolysis (500 °C, 2 h, NH_3 atm)	Water and trisodium citrate	90	6	Not given	Nanosheets with 20 nm thickness	70
$Ni(OH)_2$ on nickel foam	Growth by immersion; heat treatment (H_2 atm, 600 °C, 5 h)	Water, NaOH, $(NH_4)_2S_2O_8$	RT	6	0.01	Nanosheets	71
$ZnCo_2O_4$ on Ni foam	Polyol refluxing process; calcination (400 °C, 2 h)	Water, oxalic acid	80	2	Not given	Nanorods	72
CuO on Ni foam	Growth solution Immersion in Ni solution	Ammonia : water	90 90	2 6	7	Nanosheets with 150 nm thickness	73
β-$Ni(OH)_2$/γ-NiOOH film	Chemical bath deposition	Ammonia : water	20	1	1.4	Nanoflakes with 20 nm thickness	74
Co_3O_4 on Ni foam	Growth solution	Ammonia : water	90	2	16	Nanowires with 250 nm diameter and 15 µm length	75
Mesoporous transition metal oxides in Ni foam	Chemical bath deposition; calcination (300 °C, 1.5 h, Ar atm)	Water	85	2	Not given	Nanosheets	89

[a]pc: partially crystalline.

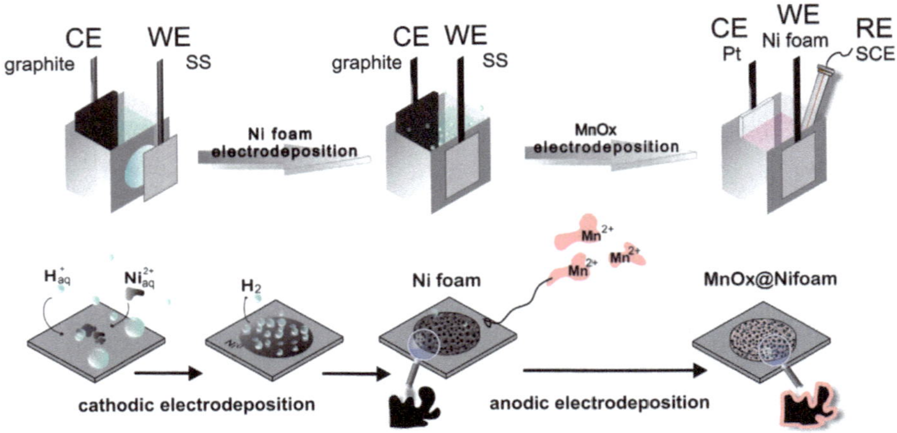

Figure 3.3 Schematic illustration of the experimental setup for the preparation of $MnO_x{\cdot}nH_2O$ foams. Reproduced from ref. 90 with permission from Elsevier, Copyright 2018.

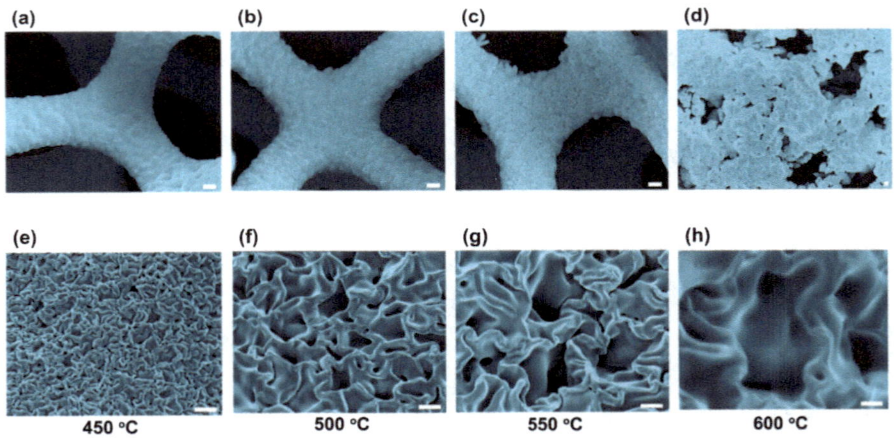

Figure 3.4 SEM images of 3D porous $NiSe_2$/Ni foam prepared at different temperatures: (a and e) 450 °C, (b and f) 500 °C, (c and g) 550 °C and (d and h) 600 °C. Scale bar: (a–d) 20 μm and (e–h) 1 μm. Reproduced from ref. 9 with permission from Elsevier, Copyright 2016.

duration of the immersion of the metal foam in the bath. The careful control of all these parameters (and others) is critical to guarantee the stability of the bare foams while achieving a rigorous control of the composition, morphology and structural features of the new material obtained.

Overall, foam functionalization is a very attractive route to create novel electrode architectures, exploring the advantages of the unique 3D properties of these metallic structures. Achieving easy functionalization, involving

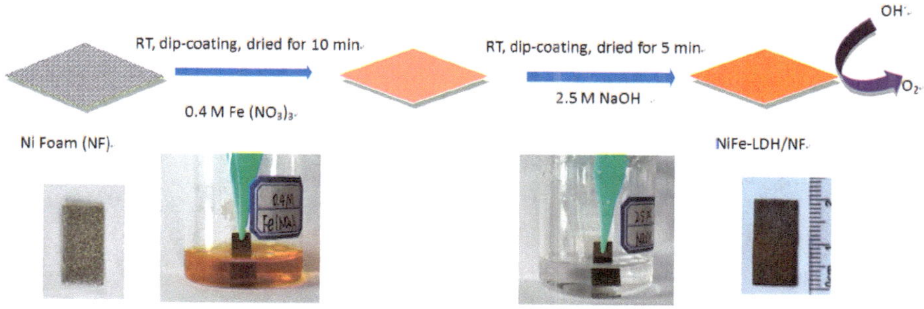

Figure 3.5 Schematic illustration of the general procedure to deposit active materials on metallic foams, by chemical bath deposition. Reproduced from ref. 76 with permission from Elsevier, Copyright 2020.

a minimum of steps while keeping the original properties of the bare foam holds promising results for industrial applications; however, there are not so many cost-effective routes available nowadays. Despite the different alternatives presented, there are still some important open challenges to overcome before reaching simple and easy to scale up functionalization routes for metal foams.

References

1. H. Sun, Z. Yan, F. Liu, W. Xu, F. Cheng and J. Chen, *Adv. Mater.*, 2020, **32**, e1806326.
2. Y. Choi, T.-Y. Ahn, E. H. Lee, T.-U. Hur, S.-H. Ha, J.-I. Lee, T. R. Park, C. H. Choi, J.-Y. Kim and J.-H. Cho, *Mater. Adv.*, 2021, **2**, 1972–1980.
3. B. A. Mola, G. Mani, M. R. Pallavolu, N. R. Reddy, N. S. Alsaiari, F. M. Alzahrani, K. M. Katubi, Y. Noh, S. K. Jilcha, H.-J. Kim and Y. A. Kumar, *J. Energy Storage*, 2021, **44**, 103417.
4. P. Hu, T. Wang, J. Zhao, C. Zhang, J. Ma, H. Du, X. Wang and G. Cui, *ACS Appl. Mater. Interfaces*, 2015, **7**, 26396–26399.
5. K. I. Siwek, S. Eugénio, T. M. Silva and M. Fatima Montemor, *Energy Technol.*, 2019, **7**, 1801139.
6. J. Cao, J. Zhou, Y. Zhang, Y. Wang and X. Liu, *ACS Appl. Mater. Interfaces*, 2018, **10**, 1752–1760.
7. J. Li, G. Wei, Y. Zhu, Y. Xi, X. Pan, Y. Ji, I. V. Zatovsky and W. Han, *J. Mater. Chem. A*, 2017, **5**, 14828–14837.
8. M. Yan, Y. Yao, J. Wen, L. Long, M. Kong, G. Zhang, X. Liao, G. Yin and Z. Huang, *ACS Appl. Mater. Interfaces*, 2016, **8**, 24525–24535.
9. H. Zhou, Y. Wang, R. He, F. Yu, J. Sun, F. Wang, Y. Lan, Z. Ren and S. Chen, *Nano Energy*, 2016, **20**, 29–36.
10. C. Xiao, Y. Li, X. Lu and C. Zhao, *Adv. Funct. Mater.*, 2016, **26**, 3515–3523.
11. K. Lota, P. Swoboda, I. Acznik, A. Sierczyńska, R. Mańczak, Ł. Kolanowski and G. Lota, *Curr. Appl. Phys.*, 2020, **20**, 106–113.

12. A. Lakshmi-Narayana, M. Dhananjaya, N. Guru-Prakash, A. Mauger, C. M. Julien and O. M. Hussain, *Heliyon*, 2019, **5**, e02060.
13. J. Lin, Y. Yan, X. Zheng, Z. Zhong, Y. Wang, J. Qi, J. Cao, W. Fei, Y. Huang and J. Feng, *J. Colloid Interface Sci.*, 2019, **536**, 456–462.
14. H. Peng, C. Jing, J. Chen, D. Jiang, X. Liu, B. Dong, F. Dong, S. Li and Y. Zhang, *CrystEngComm*, 2019, **21**, 470–477.
15. J. Lin, H. Wang, Y. Yan, X. Zheng, H. Jia, J. Qi, J. Cao, J. Tu, W. Fei and J. Feng, *J. Mater. Chem. A*, 2018, **6**, 19151–19158.
16. H. Gao, X. Wang, G. Wang, C. Hao, S. Zhou and C. Huang, *Nanoscale*, 2018, **10**, 10190–10202.
17. M. Huang, Y. Zhang, F. Li, L. Zhang, Z. Wen and Q. Liu, *J. Power Sources*, 2014, **252**, 98–106.
18. C. Guan, J. Liu, C. Cheng, H. Li, X. Li, W. Zhou, H. Zhang and H. J. Fan, *Energy Environ. Sci.*, 2011, **4**, 4496–4499.
19. H. Huo, Y. Zhao and C. Xu, *J. Mater. Chem. A*, 2014, **2**, 15111–15117.
20. J. Xu, Y. Sun, M. Lu, L. Wang, J. Zhang, J. Qian and X. Liu, *Chem. Eng. J.*, 2018, **334**, 1466–1476.
21. B. Liu, Y. F. Zhao, H. Q. Peng, Z. Y. Zhang, C. K. Sit, M. F. Yuen, T. R. Zhang, C. S. Lee and W. J. Zhang, *Adv. Mater.*, 2017, **29**, 1606521.
22. C. Du, M. Shang, J. Mao and W. Song, *J. Mater. Chem. A*, 2017, **5**, 15940–15949.
23. A. Sivanantham and S. Shanmugam, *Appl. Catal., B*, 2017, **203**, 485–493.
24. W. Zhou, X.-J. Wu, X. Cao, X. Huang, C. Tan, J. Tian, H. Liu, J. Wang and H. Zhang, *Energy Environ. Sci.*, 2013, **6**, 2921–2924.
25. J. Hu, M. Li, F. Lv, M. Yang, P. Tao, Y. Tang, H. Liu and Z. Lu, *J. Power Sources*, 2015, **294**, 120–127.
26. H. Chen, J. Jiang, L. Zhang, D. Xia, Y. Zhao, D. Guo, T. Qi and H. Wan, *J. Power Sources*, 2014, **254**, 249–257.
27. K. Xu, W. Li, Q. Liu, B. Li, X. Liu, L. An, Z. Chen, R. Zou and J. Hu, *J. Mater. Chem. A*, 2014, **2**, 4795–4802.
28. Q. Wang, X. Wang, B. Liu, G. Yu, X. Hou, D. Chen and G. Shen, *J. Mater. Chem. A*, 2013, **1**, 2468–2473.
29. D. Guo, P. Zhang, H. Zhang, X. Yu, J. Zhu, Q. Li and T. Wang, *J. Mater. Chem. A*, 2013, **1**, 9024–9027.
30. A. Sivanantham, P. Ganesan and S. Shanmugam, *Adv. Funct. Mater.*, 2016, **26**, 4661–4672.
31. F. Zhang, C. Yuan, X. Lu, L. Zhang, Q. Che and X. Zhang, *J. Power Sources*, 2012, **203**, 250–256.
32. M. Zhang, T. Wang, H. Cao, S. Cui and P. Du, *J. Energy Chem.*, 2020, **42**, 71–76.
33. A. M. Zardkhoshoui and S. S. H. Davarani, *J. Colloid Interface Sci.*, 2019, **535**, 195–204.
34. K.-L. Yan, J.-F. Qin, Z.-Z. Liu, B. Dong, J.-Q. Chi, W.-K. Gao, J.-H. Lin, Y.-M. Chai and C.-G. Liu, *Chem. Eng. J.*, 2018, **334**, 922–931.
35. F. Sun, G. Wang, Y. Ding, C. Wang, B. Yuan and Y. Lin, *Adv. Energy Mater.*, 2018, **8**, 1800584.

36. S. Shit, S. Chhetri, W. Jang, N. C. Murmu, H. Koo, P. Samanta and T. Kuila, *ACS Appl. Mater. Interfaces*, 2018, **10**, 27712–27722.
37. D. Cai, D. Wang, C. Wang, B. Liu, L. Wang, Y. Liu, Q. Li and T. Wang, *Electrochim. Acta*, 2015, **151**, 35–41.
38. C. Ouyang, X. Wang, C. Wang, X. Zhang, J. Wu, Z. Ma, S. Dou and S. Wang, *Electrochim. Acta*, 2015, **174**, 297–301.
39. J. Liu, J. Wang, B. Zhang, Y. Ruan, L. Lv, X. Ji, K. Xu, L. Miao and J. Jiang, *ACS Appl. Mater. Interfaces*, 2017, **9**, 15364–15372.
40. C. Zhang, X. Geng, S. Tang, M. Deng and Y. Du, *J. Mater. Chem. A*, 2017, **5**, 5912–5919.
41. X. Chen, D. Chen, X. Guo, R. Wang and H. Zhang, *ACS Appl. Mater. Interfaces*, 2017, **9**, 18774–18781.
42. G. K. Veerasubramani, K. Krishnamoorthy and S. J. Kim, *J. Power Sources*, 2016, **306**, 378–386.
43. R. Li, S. Wang, Z. Huang, F. Lu and T. He, *J. Power Sources*, 2016, **312**, 156–164.
44. Z. Zhang, H. Zhang, X. Zhang, D. Yu, Y. Ji, Q. Sun, Y. Wang and X. Liu, *J. Mater. Chem. A*, 2016, **4**, 18578–18584.
45. K. N. Hui, K. S. Hui, Z. Tang, V. V. Jadhav and Q. X. Xia, *J. Power Sources*, 2016, **330**, 195–203.
46. J. Pu, T. Wang, H. Wang, Y. Tong, C. Lu, W. Kong and Z. Wang, *ChemPlusChem*, 2014, **79**, 577–583.
47. C. Tang, N. Cheng, Z. Pu, W. Xing and X. Sun, *Angew. Chem., Int. Ed.*, 2015, **54**, 9351–9355.
48. W. Kong, C. Lu, W. Zhang, J. Pu and Z. Wang, *J. Mater. Chem. A*, 2015, **3**, 12452–12460.
49. M. Yu, J. Chen, J. Liu, S. Li, Y. Ma, J. Zhang and J. An, *Electrochim. Acta*, 2015, **151**, 99–108.
50. Z. Zhang, Q. Wang, C. Zhao, S. Min and X. Qian, *ACS Appl. Mater. Interfaces*, 2015, **7**, 4861–4868.
51. W. Hu, R. Chen, W. Xie, L. Zou, N. Qin and D. Bao, *ACS Appl. Mater. Interfaces*, 2014, **6**, 19318–19326.
52. L. Yu, G. Zhang, C. Yuan and X. W. Lou, *Chem. Commun.*, 2013, **49**, 137–139.
53. D. Su, Z. Tang, J. Xie, Z. Bian, J. Zhang, D. Yang, D. Zhang, J. Wang, Y. Liu, A. Yuan and Q. Kong, *Appl. Surf. Sci.*, 2019, **469**, 487–494.
54. Q. He, T. Yang, X. Wang, P. Zhou, S. Chen, F. Xiao, P. He, L. Jia, T. Zhang and D. Yang, *J. Mater. Sci.: Mater. Electron.*, 2021, **32**, 3649–3660.
55. V. H. Nguyen and J.-J. Shim, *J. Power Sources*, 2015, **273**, 110–117.
56. C. Yuan, J. Li, L. Hou, X. Zhang, L. Shen and X. W. D. Lou, *Adv. Funct. Mater.*, 2012, **22**, 4592–4597.
57. C. Yuan, L. Yang, L. Hou, L. Shen, X. Zhang and X. W. Lou, *Energy Environ. Sci.*, 2012, **5**, 7883–7887.
58. Y. Li, S. Chang, X. Liu, J. Huang, J. Yin, G. Wang and D. Cao, *Electrochim. Acta*, 2012, **85**, 393–398.

59. C. Zhao, X. Wang, S. Wang, Y. Wang, Y. Zhao and W. Zheng, *Int. J. Hydrogen Energy*, 2012, **37**, 11846–11852.
60. L.-B. Kong, M.-C. Liu, J.-W. Lang, M. Liu, Y.-C. Luo and L. Kang, *J. Solid State Electrochem.*, 2010, **15**, 571–577.
61. S. Feng, *Int. J. Electrochem. Sci.*, 2020, 2806–2821.
62. X. Chen, Y. Meng, T. Gao, J. Zhang, X. Li, H. Yuan and D. Xiao, *Sustainable Energy Fuels*, 2020, **4**, 331–336.
63. H. Zhou, F. Yu, J. Sun, R. He, S. Chen, C. W. Chu and Z. Ren, *Proc. Natl. Acad. Sci. U. S. A.*, 2017, **114**, 5607–5611.
64. W. Zhu, X. Yue, W. Zhang, S. Yu, Y. Zhang, J. Wang and J. Wang, *Chem. Commun.*, 2016, **52**, 1486–1489.
65. Z. Zhang, J. Wang, X. Yan, S. Zhang, W. Yang, Z. Zhuang and W.-Q. Han, *Energy Storage Mater.*, 2020, **29**, 332–340.
66. M. Guo, A. Qayum, S. Dong, X. Jiao, D. Chen and T. Wang, *J. Mater. Chem. A*, 2020, **8**, 9239–9247.
67. U. Y. Qazi, R. Javaid, N. Tahir, A. Jamil and A. Afzal, *Int. J. Hydrogen Energy*, 2020, **45**, 33396–33406.
68. Y. X. Zhu, L. Zhang, G. G. Zhu, X. Zhang and S. Y. Lu, *J. Colloid Interface Sci.*, 2020, **562**, 42–51.
69. X. L. Guo, X. Y. Liu, X. D. Hao, S. J. Zhu, F. Dong, Z. Q. Wen and Y. X. Zhang, *Electrochim. Acta*, 2016, **194**, 179–186.
70. B. Zhang, C. Xiao, S. Xie, J. Liang, X. Chen and Y. Tang, *Chem. Mater.*, 2016, **28**, 6934–6941.
71. X. Xiong, D. Ding, D. Chen, G. Waller, Y. Bu, Z. Wang and M. Liu, *Nano Energy*, 2015, **11**, 154–161.
72. B. Liu, B. Liu, Q. Wang, X. Wang, Q. Xiang, D. Chen and G. Shen, *ACS Appl. Mater. Interfaces*, 2013, **5**, 10011–10017.
73. G. Wang, J. Huang, S. Chen, Y. Gao and D. Cao, *J. Power Sources*, 2011, **196**, 5756–5760.
74. Y. F. Yuan, X. H. Xia, J. B. Wu, J. L. Yang, Y. B. Chen and S. Y. Guo, *Electrochim. Acta*, 2011, **56**, 2627–2632.
75. Y. Gao, S. Chen, D. Cao, G. Wang and J. Yin, *J. Power Sources*, 2010, **195**, 1757–1760.
76. H. Li, X. Wang, T. Wang and F. Xiao, *J. Alloys Compd.*, 2020, **844**, 156224.
77. G. Zhu, X. Li, Y. Liu, Y. Mao, Z. Liang, Z. Ji, X. Shen, J. Sun, X. Cheng and J. Mao, *J. Mater. Chem. A*, 2020, **8**, 12603–12612.
78. A. Rabenau, *Angew. Chem., Int. Ed. Engl.*, 1985, **24**, 1026–1040.
79. D. O'Hare, Hydrothermal Synthesis, *Encyclopedia of Materials: Science and Technology*, 2nd edn, 2001, pp. 3989–3992.
80. S. Feng and L. Guanghua, in *Modern Inorganic Synthetic Chemistry*, ed. R. Xu, W. Pang and Q. Huo, Elsevier B.V., 2011, ch. 4, pp. 63–95.
81. B. D. Anderson and J. B. Tracy, *Nanoscale*, 2014, **6**, 12195–12216.
82. G. Yilmaz and X. Lu, *ChemNanoMat*, 2016, **2**, 719–725.
83. R. Liu, R. Jiang, Y. H. Chu and W. D. Yang, *Nanomaterials*, 2021, **11**, 2736.
84. S. Yang, B. Deng, R. Ge, L. Zhang, H. Wang, Z. Zhang, W. Zhu and G. Wang, *Nanoscale Res. Lett.*, 2014, **9**, 672.

85. T. Nguyen, S. Eugenio, M. Boudard, L. Rapenne, M. J. Carmezim, T. M. Silva and M. F. Montemor, *Nanoscale*, 2015, 7, 12452–12459.
86. K. I. Siwek, S. Eugénio, I. Aldama, J. M. Rojo, J. M. Amarilla, A. P. C. Ribeiro, T. M. Silva and M. F. Montemor, *J. Electrochem. Soc.*, 2022, **169**, 020511.
87. D. H. Nam, R. H. Kim, D. W. Han and H. S. Kwon, *Electrochim. Acta*, 2012, **66**, 126–132.
88. M. Z. Ansari, N. Parveen, D. K. Nandi, R. Ramesh, S. A. Ansari, T. Cheon and S. H. Kim, *Sci. Rep.*, 2019, **9**, 10225.
89. Q. Zhang, J. Wang, J. Dong, F. Ding, X. Li, B. Zhang, S. Yang and K. Zhang, *Nano Energy*, 2015, **13**, 77–91.
90. K. I. Siwek, S. Eugénio, T. M. Silva and M. F. Montemor, *Electrochim. Acta*, 2018, **281**, 39–47.
91. F. Zhang, Y. Ge, H. Chu, P. Dong, R. Baines, Y. Pei, M. Ye and J. Shen, *ACS Appl. Mater. Interfaces*, 2018, **10**, 7087–7095.
92. J. Yan, L. Chen and X. Liang, *Sci. Bull.*, 2019, **64**, 158–165.
93. A. Muthurasu, B. Dahal, K. Chhetri and H. Y. Kim, *Inorg. Chem.*, 2020, **59**, 3817–3827.
94. Q. Yuan, Y. Yu, Y. Gong and X. Bi, *ACS Appl. Mater. Interfaces*, 2020, **12**, 3592–3602.
95. F. Miao, X. Li, B. Tao and P. K. Chu, *Ionics*, 2020, **26**, 5241–5249.

CHAPTER 4

Property Characterization of Metal Foams

4.1 Specific Surface Area

In the field of electrocatalysis and electrochemical energy storage, the specific surface area (SSA) is a paramount feature. Larger surface areas notably increase the number of active sites where electrochemical reactions and other surface (or near-surface) processes occur. In electrochemical devices, a high mass loading of active material per volume is required to achieve the desired content of material and electron transport properties. This would be a challenge without the use of 3D architectures, such as metallic foams, that are excellent enablers to reach optimal material loadings.[1] Some of the nanostructured materials that show a large active surface area are porous carbon-based materials (*e.g.*, activated carbons, carbide derived carbons and carbon nanotubes), polymers, whose selection must be based on their other intrinsic properties, and foam-based materials.[2,3]

Structural properties such as porosity and roughness are contributing factors that determine the SSA of a given material. On the one hand, porosity is described as the volumetric fraction of empty space in the full volume of the material, and it can be fine-tuned by controlling the synthesis processes,[4] as discussed in Chapter 2. An important metric is 'porosity fraction' and it is defined as the ratio between volumes: V_m is the volume effectively occupied

Transition Metal-based Nanofoams for Electrochemical Systems: Manufacturing, Characterization and Applications
By Maria de Fátima Montemor, Diana M. Fernandes, Alberto Adan-Mas, Ana Catarina Alves, Gabriel Garcia Carvalho and Inês S. Marques

Published by the Royal Society of Chemistry, www.rsc.org

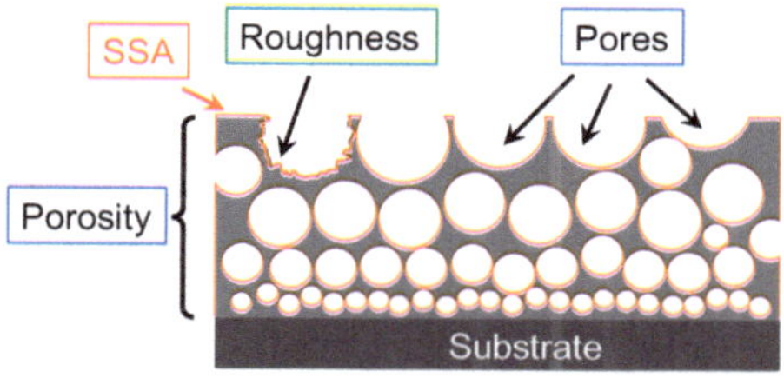

Figure 4.1 Representation of pore structure, roughness and specific surface area in a porous material.

by the material, V_T is the total volume and V_p is the volume occupied by pores, according to eqn (4.1):

$$\varepsilon = \frac{V_T - V_m}{V_T} = \frac{V_p}{V_T} \tag{4.1}$$

On the other hand, roughness refers to the texture of the surface of the material and represents the directional deviations from the surface's ideal plane, increasing the active sites for the same mass loading of material.[1] Due to the highly complex and ultimately random definition of surfaces, roughness is often quantified by geometric amplitude parameters based on statistical methods. An in-depth review about surface roughness measurements was reported by David Whitehouse.[5] Figure 4.1 schematically shows the relationship between SSA, roughness and pores and how these properties constitute a better understanding of the surface area in metal foams, assuming circular pores.

Depending on their size, pores can be classified in three main categories: micropores (<2 nm), mesopores (2 to 50 nm) and macropores (>50 nm). The characterization of the surface features of porous materials is vital to understanding the correlation between composition, morphology and performance. For this reason, multiple techniques have been developed to gain in-depth knowledge on the micro- and nanostructure of these materials, particularly adsorption/desorption, electrochemical active surface area and imaging techniques. An overview of these methods is described in the following subsections.

4.1.1 Adsorption/Desorption Techniques: Brunauer–Emmett–Teller

The quantification of the SSA of metal foams is usually performed by means of physical adsorption/desorption processes, also known as physisorption. In this process, the surface of the material is accessed by an adsorbate that covers it. Adsorption/desorption processes are usually studied using the Brunauer–Emmett–Teller (BET) and Barrett–Joyner–Halenda (BJH) theories, which will be presented in this subsection. The BET theory aims to define the adsorption and desorption processes of inert gases (hereby named as adsorbate) onto a porous structure (hereby named as adsorbent). It is an experimental technique used to determine the specific surface area (SSA) of a material by establishing mathematical relationships between its thermodynamic properties in an adsorption/desorption system. The value obtained from this method is typically referred to as S_{BET}. A thorough examination and exemplification of BET equations is described by Raja *et al.*[6]

The determination of S_{BET} is based on the creation of a monolayer of the adsorbate on top of the porous material which fully covers the exposed surface, as depicted in Figure 4.2. By measuring the adsorbed volume of molecules during the increase of the relative pressure of the adsorbate (in relation to its vapour pressure at the same temperature), experimental adsorption isotherm curves can be obtained and analysed towards the understanding of the adsorption relationship between adsorbate and adsorbent. The inverse process can also be analysed in the form of desorption curves obtained for a decrease of relative pressure. The corresponding curve pairs are often named adsorption–desorption curves or hysteresis loop depending on the level of overlap that the curves showcase. The shape of these curves is classified by an IUPAC definition, usually conforming to type II and IV(a) isotherms.[7] The isotherms typically showcase a plateau, which indicates the formation of monolayers over the mesopores. This plateau is then interrupted by a steep increase of adsorbed volume at higher relative pressures at which point multilayers are being formed on the mesopores as the pores are fully occupied until the test ends. Depending on the type of metal foam being studied, the

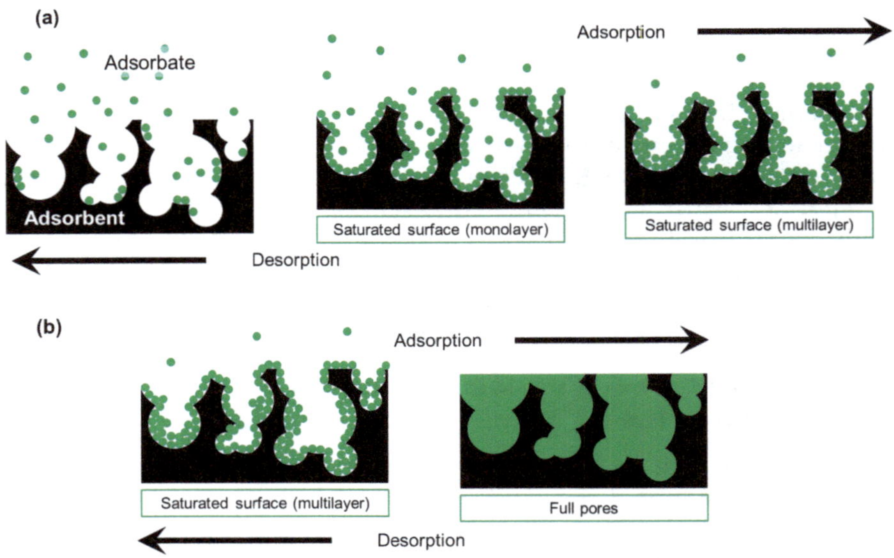

Figure 4.2 Schematic representation of the (a) three main stages of adsorption/desorption curves from BET analysis; and (b) pore filling of BJH.

existence of macropores can also be detected if a second plateau is observed. The shape of the pores also influences the shape of the isotherms as different synthesis methodologies can produce metal foams with very distinctive empty spaces (*e.g.*, pores or voids).[8,9] Therefore, BET methodology is used for several analyses of metal foams.

Lee *et al.*[10] used BET analysis to study different metal/carbon foam composites, observing a type II curve that commonly appears in micropores' structures, as shown in Figure 4.3(a). Moreover, the flat isotherm in the range of 0.2 to 0.8 $[P/P_0]$ was indicative of few mesopores in the foams. The different curves of the foams reflected various SSA values, in which the sample containing 15% of activated carbon revealed a higher value (106.48 $m^2\ g^{-1}$) compared to the plain foam without carbon material (10.75 $m^2\ g^{-1}$).[10] Another example is reported by Gao *et al.*,[11] where BET analysis was used to evaluate the porosity of different open-cell foams prepared by combustion synthesis using different concentrations of Ni^{2+} (ranging from 0.35 to 4.0 M). With the increased concentration of the Ni^{2+}-based precursor, the BET surface-area of the foams decreased from 6.5 to 2.1 $m^2\ g^{-1}$, as demonstrated by the isotherm IV type with hysteresis loops in Figure 4.3(b).[11] The experimental data allow understanding of the types of pores present in the structure by analysing the shape of the resulting isotherms, which will depend on the behaviour of the adsorbate within the pores of the adsorbent and the available SSA.[12]

Often complementary to BET analysis, the Barrett–Joyner–Halenda (BJH) method is commonly represented alongside the adsorption–desorption isotherms to assess the pore size distribution. The BJH method is applied when the adsorbent is completely saturated with adsorbate, as depicted in Figure 4.2(b). In that stage of adsorption, the pore size distribution of porous materials can be analysed from their experimental isotherms under

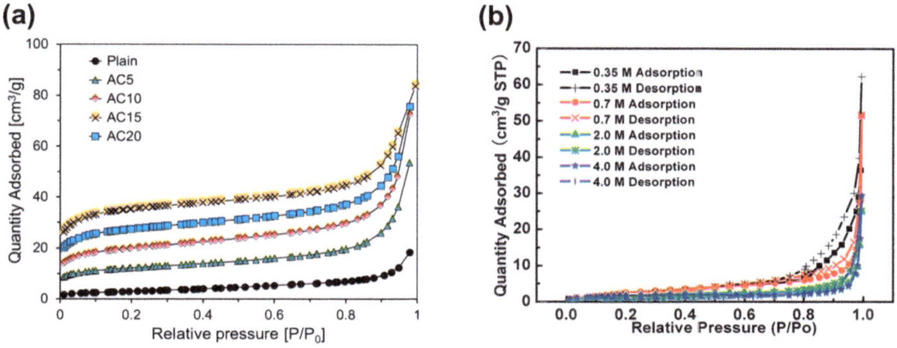

Figure 4.3 Examples of isotherm curves of metal foams: (a) metal/carbon composite curves. Reproduced from ref. 10, https://doi.org/10.3390/ma13184038, under the terms of the CC BY 4.0 license, http://creativecommons.org/licenses/by/4.0/. (b) Ni foams produced at different concentrations. Reproduced from ref. 11 with permission from the Royal Society of Chemistry.

the application of the Kelvin model for pore filling.[13] Briefly, as the material is completely covered and its pores are full of adsorbate, the Kelvin model assumes capillary conditions and the presence of an adsorbate film with a thickness that can be determined as a function of the relative pressure decrease during desorption.[14,15] Hence, the calculated thickness is processed by the model to determine the pore size. An example of applying the BJH method is reported in the work of Li *et al.*,[16] who determined the distribution of pore sizes of three Ni–Co oxide metal foams with different composition ratios of Ni and Co, as well as the average pore size for each composite. The inset graphs of Figure 4.4(a–c) show the pore size distributions from experimental isotherms using the Kelvin model of pore filling of the Ni–Co oxide metal foams.

When applying mathematical models to describe physisorption processes, the BET theory and BJH model for pore filling have limited applicability. The most notable limitations to these techniques are: (1) the natural existence of other adsorbate–adsorbent interactions; (2) the heterogeneity of the surface; (3) the thermodynamic variation of the adsorption energies, namely the heat of adsorption; (4) the presence of closed pores unable to be accessed

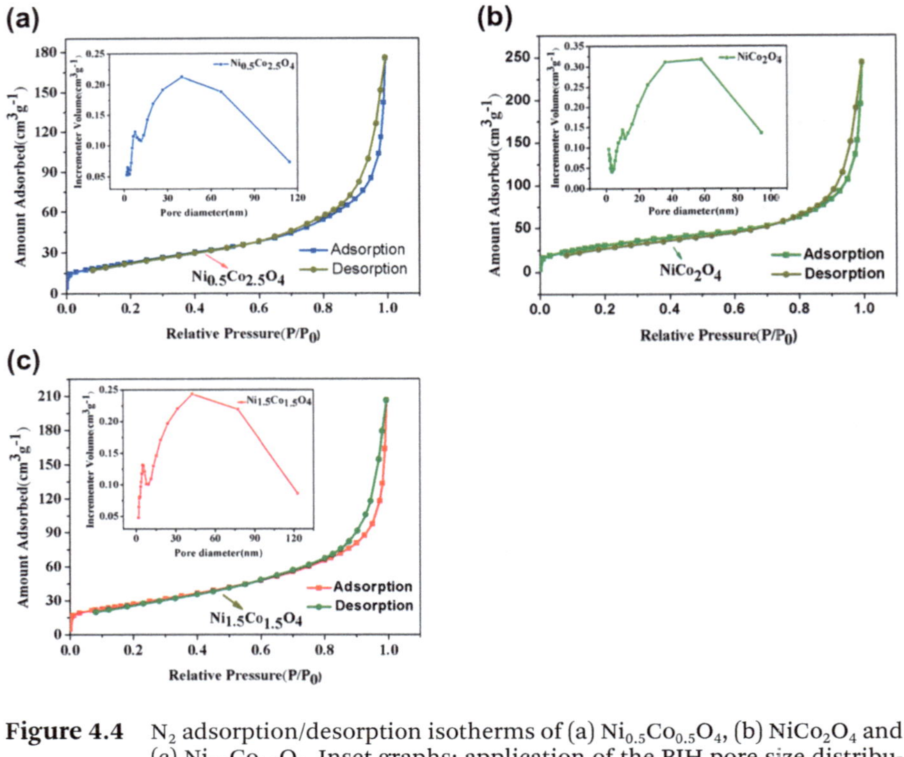

Figure 4.4 N_2 adsorption/desorption isotherms of (a) $Ni_{0.5}Co_{0.5}O_4$, (b) $NiCo_2O_4$ and (c) $Ni_{1.5}Co_{1.5}O_4$. Inset graphs: application of the BJH pore size distribution. Reproduced from ref. 16 with permission from Springer Nature, Copyright 2019.

by the adsorbate; and, in certain cases, (5) the challenging sample preparation.[17] Nevertheless, both BET and BJH techniques allow a complete and reliable quantification of the porous structure of metal foams in the same experimental setup. Furthermore, recent studies demonstrated that the SSA of different metal foams can be modelled and calculated computationally with a high level of precision when compared to BET and BJH experimental determinations. To overcome the many limitations of these techniques, other advanced characterization techniques have been developed, such as the Dubinin–Radushkevich (DR) equation,[18] semi-experimental models such as the Horvath–Kawazoe (HK) method,[19] and computational DFT-based models (*e.g.*, non-local density functional theory, NLDFT); these can be found in the literature[20–22] and may be applied to metal foams.

4.1.2 Electrochemical Methods

One of the most interesting characteristics of metal foams for electrochemical applications is their electrochemically active surface (ECSA) which affects non-faradaic electrostatic interactions and faradaic redox reactions. It is possible to determine the ECSA of metal foams by using an adequate potential range in an appropriate electrolyte solution. While the SSA represents the area per unit of mass, the ECSA defines the area accessible by an electrolyte and capable of hosting double layer or charge transfer processes for the same unit of mass.[23]

Electrochemical tests, namely cyclic voltammetry (CV), are used to measure current *vs.* potential dependence, and to assess the regions where there is a purely double layer/electrostatic contribution to determine the ECSA. Experimentally, Rodríguez *et al.*[23] determined the real surface area of platinum electrodes in acidic electrolytes by CV shown in Figure 4.5, where

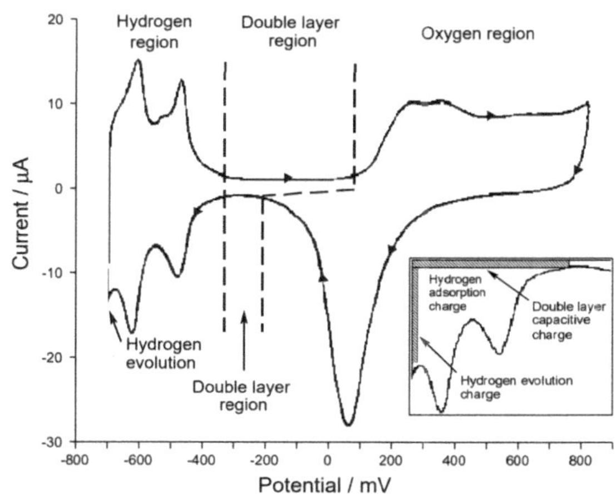

Figure 4.5 Cyclic voltammetric current potential curve for a platinum electrode in 0.5 M H_2SO_4 solution. Reproduced from ref. 23 with permission from American Chemical Society, Copyright 2000.

three regions were highlighted by the author: the hydrogen evolution region, the double-layer region and the oxygen evolution region. Other techniques focused on the determination of the real surface area of certain materials include CO stripping and underpotential deposition.

4.1.2.1 Cyclic Voltammetry: Double-layer Region

The double-layer region, also known as the capacitive region, represents the potential window at which only the electrochemical double-layer (EDL) is present. The EDL is governed by a charge separation mechanism that stores electrostatic charge with both specific and non-specific adsorption of electrolyte ions at the electrode–electrolyte interface.[24] In this case, the ECSA is determined by analysing the data obtained from the capacitive region of the cyclic voltammogram of the sample. This region can be identified in a wide potential window cyclic voltammogram, since it does not present any relevant or noticeable current peaks, *i.e.*, the faradaic response of the system is negligible. In this method, the potential window of the CV is narrowed to the capacitive region, as shown in Figure 4.6(a), followed by a kinetic study of the CV at different scan rates (Figure 4.6(b)). After fixing a potential value, the corresponding cathodic average currents are plotted as a function of the scan rate, resulting in a linear (Figure 4.6(c)) or quadratic curve, depending on the capacitive or pseudocapacitive behaviour of the material. Assuming a capacitive trend, the linear curve shows a slope that corresponds to the value of the double-layer capacitance, C_{DL}, that is determined by the following equations:[25]

$$C_{DL} = \frac{dQ}{d\varphi} \tag{4.2}$$

$$i = \frac{dQ(\varphi)}{dt} \tag{4.3}$$

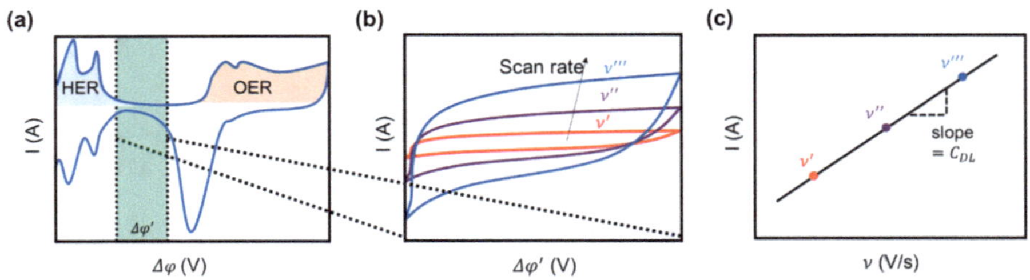

Figure 4.6 Schematic representation of the electrochemical method for surface area determination with the double-layer region: (a) definition of capacitive region $\Delta\varphi'$ and identification of the hydrogen evolution reaction (HER) and oxygen evolution reaction (OER); (b) CVs at different scan rates for $\Delta\varphi'$; and (c) typical capacitive linear relation of cathodic average currents as a function of scan rates (ν) at fixed potential.

in which Q is the charge and φ is the potential. From the combination of eqn (4.2) and (4.3), the mathematical expression that correlates the current and the scan rate ($v = \mathrm{d}\varphi/\mathrm{d}t$) is obtained, as shown in Figure 4.6(c):

$$i = C_{\mathrm{DL}} \frac{\mathrm{d}\varphi}{\mathrm{d}t} \tag{4.4}$$

With the theoretical C_{DL} value, it is possible to determine the ECSA using eqn (4.5).

$$\mathrm{ECSA} = \frac{C_{\mathrm{DL}}}{C_{\mathrm{S}}} \tag{4.5}$$

where C_{S} is the specific capacitance of the material that is usually a standard value for alkaline or acid solutions.

From the three possibilities, the double-layer region is the most used region in a cyclic voltammogram to determine the ECSA of metal foams, while the hydrogen and oxygen regions are often associated with the study of electrocatalytic activity. In the work of Siwek *et al.*,[26] different Ni foams prepared using the DHBT technique were studied to reveal their double-layer region and roughness factor, that, in this case, was equal to the ECSA. Figure 4.7 shows the cyclic voltammograms of a Ni foam, as well as a close-up view of the double-layer region, and the linear slope of the capacitive current as a function of the scan rate for three different foams. The ECSA value is directly related to the double-layer capacitance of the foam. The slope shown in Figure 4.7(c) demonstrated a higher ECSA value for a 3D Ni foam, with a distinct surface heterogeneity, low pore distribution and high deposited mass, resulting in a high capacitance value (12.8 mF cm^{-2}).[26] Compared to the other Ni foams, distinct surface heterogeneity, lower pore distribution and higher deposited mass on the substrate enabled the increase of the ECSA.[26]

Diao *et al.*[27] determined the ECSA of an open-cell Cu porous material whose voltammogram consisted of three anodic peaks corresponding to different electrochemical reactions marked in the cyclic voltammogram on Figure 4.8(a):

$$2\mathrm{Cu} + \mathrm{OH}^- \rightarrow \mathrm{Cu_2O} + \mathrm{H_2O} + 2\mathrm{e} \quad (\text{peak I})$$

$$\mathrm{Cu_2O} + 2\mathrm{OH}^- + \mathrm{H_2O} \rightarrow 2\mathrm{Cu(OH)_2} + 2\mathrm{e} \quad (\text{peak II})$$

$$\mathrm{Cu} + 2\mathrm{OH}^- \rightarrow \mathrm{CuO} + \mathrm{H_2} + 2\mathrm{e} \quad (\text{peak III})$$

Although the cyclic voltammogram of the porous Cu showed sharp redox peaks, in the potential range of −1 to −0.75 V, no faradaic current was observed, and hence, the double-layer capacitance was determined in this region (Figure 4.8).

Other methods use either the HER or OER regions of the cyclic voltammogram to determine the surface area of materials. For instance, Rodríguez *et al.*[23] combined HER at negative potentials and its adsorption on the active sites of the foam as a basis for the surface area determination. This method

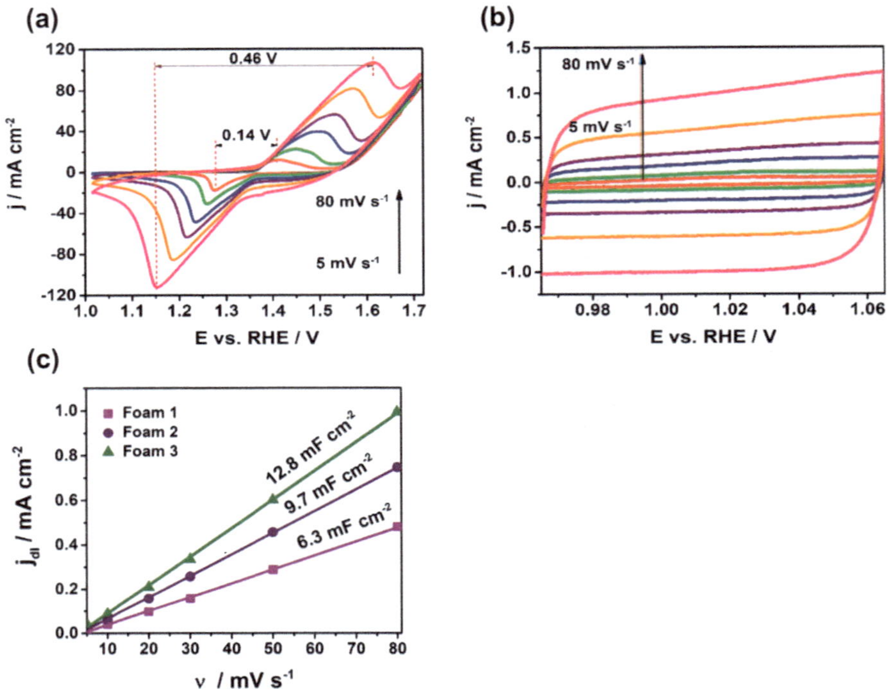

Figure 4.7 Determination of the ECSA: (a) cyclic voltammogram of different 3D Ni foams recorded at 5, 10, 20, 30, 50 and 80 mV s^{-1} (foams 1, 2 and 3 differ in the electrolyte composition used for their electrodeposition); (b) double-layer capacitance measurements for Ni foam 3 recorded at the 5–80 mV s^{-1} range; and (c) charging current density differences plotted as a function of the potential scan rate. Reproduced from ref. 26 with permission from Elsevier, Copyright 2019.

relies on the quantification of the amount of adsorbed gas in the material and assumes that one hydrogen molecule adsorbs to one molecule of surface metal. The area under the peak area that corresponds to the hydrogen region allows quantification of the charge of the monolayer (Q_m) that was adsorbed during HER.[23] The number of adsorbed hydrogen molecules, n_a, can be then calculated with the Faraday constant, F, as stated by eqn (4.6):

$$n_a = \frac{Q_m}{F} \tag{4.6}$$

which can then be used to calculate the ECSA using eqn (4.7):

$$\text{ECSA} = \frac{n_a N_A}{d_m} \tag{4.7}$$

where N_A is Avogadro's number and d_m the surface density of a metal atom.

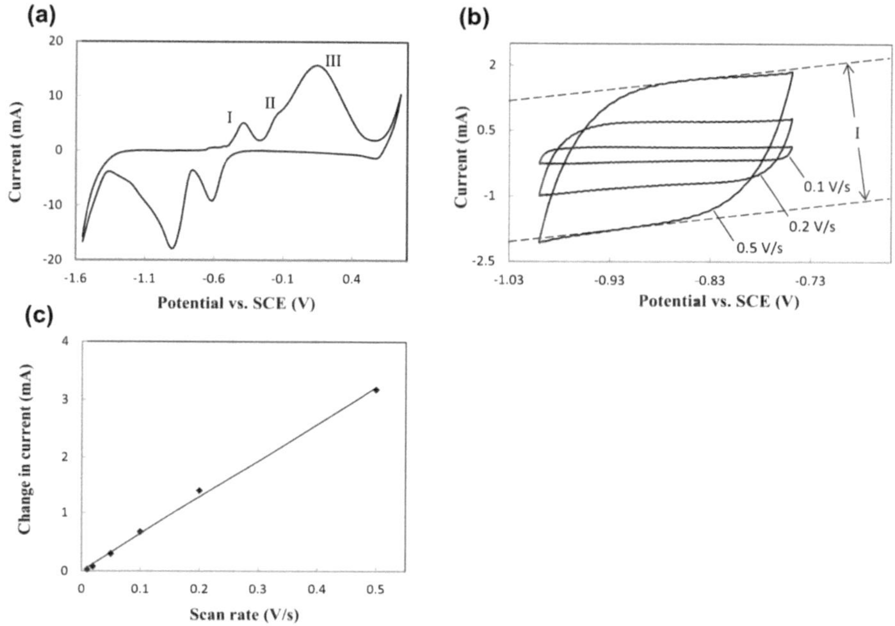

Figure 4.8 (a) Cyclic voltammogram of a typical porous Cu sample; (b) double-layer capacitance region with different scan rates; and (c) linear relation between change in current and scan rate. Reproduced from ref. 27 with permission from Elsevier, Copyright 2015.

On the other hand, at positive potentials, oxygen adsorption is used as adsorbent, employing the oxygen evolution reaction at the surface of the electrode. This reaction can happen in either acidic or alkaline media, according to the following reactions:

$$2H_2O \rightarrow O_2 + 4H^+ + 4e^- \quad (\text{acidic media})$$

$$4OH^- \rightarrow 2H_2O + O_2 + 4e^- \quad (\text{alkaline media})$$

Due to kinetic limitations related to the four electrons transferred during the generation of O_2, OER is a more complex reaction that requires multiple steps. A few possible mechanisms have been used to explain this multi-step reaction as exposed by the work of Wang *et al.*[28] for alkaline media; however, it has not been extensively used to determine the surface area.[29–31]

Although the ECSA is easily applied, one of the main limitations is that its calculous is usually an approximation. Results usually depend on inaccurate estimations of specific capacitance values taken from different bibliographic sources. Hence, this electrochemical method is often used in combination with other methods, such as BET.[25] Nevertheless, the ECSA is regularly used in the literature to understand the correlation between the electrochemical performance of materials and their surface area.

4.1.2.2 Carbon Monoxide Stripping

Carbon monoxide stripping (CO stripping) is an electrochemical method used to calculate the surface area by oxidation of previously adsorbed CO in the material using cyclic voltammetry. Due to its affinity to metals, CO is strongly adsorbed onto a metal substrate. Consequently, cyclic voltammetry is used to induce the oxidation of CO to CO_2 and the absorbent is subsequently removed from the metal surface,[32,33] resulting in a similar methodology to BET. The extent of this oxidation reaction is experimentally observed through well-defined peaks in the cyclic voltammogram, as shown in Figure 4.9.[34] The ECSA is calculated by the same logic presented before for hydrogen adsorption.

Figure 4.9(a) displays the oxidation charge of CO stripping conducted by Binninger *et al.*[35] The area highlighted beneath the curve is the charge required to complete the oxidation, Q_m, which is proportional to the amount of adsorbed gas, and, ultimately, to the ECSA. Rudi *et al.*[34] investigated the combination of CO stripping with underpotential deposition (Section 4.1.2.3) to determine the ECSA and concluded that the ECSA obtained by CO stripping was greater than that obtained by underpotential deposition. Furthermore, it was also concluded that a combined method seems more accurate for surface area characterization (Figure 4.9(b)).[34] CO stripping is well reported for fuel cell Pt-based electrocatalysts[36] and it may also be theoretically applicable for determining the surface area of metal foams. However, this technique has only been used in the scope of metal foams with very specific 3D architectures, such as metals combined with carbon nanotubes.[37]

4.1.2.3 Underpotential Deposition

Commonly associated with CO stripping, underpotential deposition (UPD) is another electrochemical process that is effectively used to determine the ECSA. The process consists of the deposition of an alien metal (or, in some

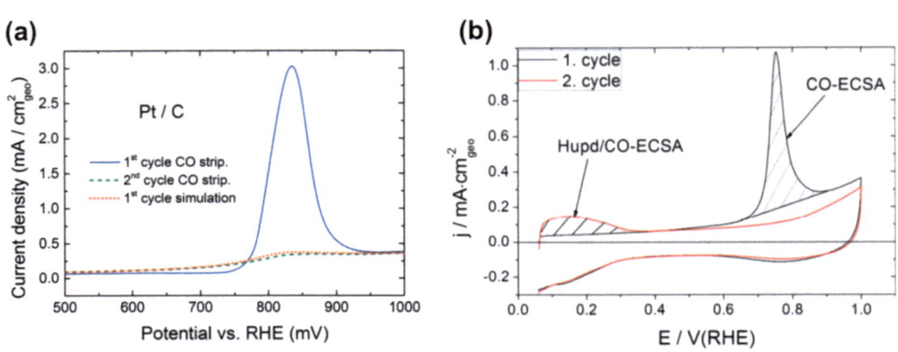

Figure 4.9 CO stripping of (a) a Pt/C catalyst in 0.1 M $HClO_4$ at 20 mV s^{-1} and (b) Pt alloy nanoparticles. (a) Reproduced from ref. 35, https://doi.org/10.1149/2.055403jes, under the terms of the CC BY 4.0 License, http://creativecommons.org/licenses/by/4.0/. (b) Reproduced from ref. 34 with permission from Springer Nature, Copyright 2014.

cases, hydrogen) onto a substrate, such as Pb on Ag[38] or Cu on Au.[39] This deposition occurs at a higher potential than the potential of reduction, forming a monolayer of the metal instead of a bulk deposition.[40] ECSA determination is then calculated from the voltametric peak area correspondent to the potentials of the UPD process.[29]

Although UPD is commonly associated with the deposition of metals, it has also been referred to as the deposition of other species and used in the determination of the ECSA. Such is the case in the aforementioned work of Rudi *et al.*[34] (Section 4.1.2.2). Notwithstanding, although theoretically feasible, neither UPD nor CO stripping are commonly used for estimating the ECSA of metal foam structures and are only suitable as complementary techniques to other surface methods. Still, some literature reports the use of UPD in metal-based foams. For example, Li *et al.*[41] studied the UPD of hydrogen onto a Pt-coated Ni foam, while Xu *et al.*[42] carried out the deposition of metals onto other 3D structures, namely the UPD of Cu single-atom catalysts onto S-doped graphite foams. Nonetheless, achieving a successful monolayer covering the entire surface of the metal foam is challenging and, considering the cost associated with using metal instead of gases, these techniques are not so widely applied. Nonetheless, this method presents an opportunity to explore alternatives where other methods fail, as well as proving useful when a metallic monolayer shall be deposited regardless or when hydrogen and oxygen regions overlap in a given CV.[29] More about the combination of CO stripping and UPD can be found in ref. 43 and 44.

4.1.3 Imaging Techniques

The morphology of metal foams shall be studied to correlate the surface morphology of the structure to their electrochemical activity. To this end, imaging techniques are used, particularly atomic force microscopy, optical and electron microscopy methods, and will be discussed in the following subsections.

4.1.3.1 Atomic Force Microscopy

Atomic force microscopy (AFM) is a high-resolution imaging technique based on a cantilever-feedback working principle. This method is able to produce images with resolutions 1000 times higher than conventional optical microscopy.[45] The AFM technique measures the force between the tip and the surface of the sample and, depending on the repulsive/attractive atomic forces and van der Waals interactions, the mode of operation of the AFM is adjusted. The main modes of operations, exemplified in Figure 4.10, are static or contact, non-contact, and intermittent or tapping AFM modes, although there are other variations.[46]

- *Contact AFM mode*

  Imaging techniques based on contact methods involve the physical interaction between the tip and the surface of the sample, which offer extremely high-resolution images.[46] For instance, Park *et al.*[47]

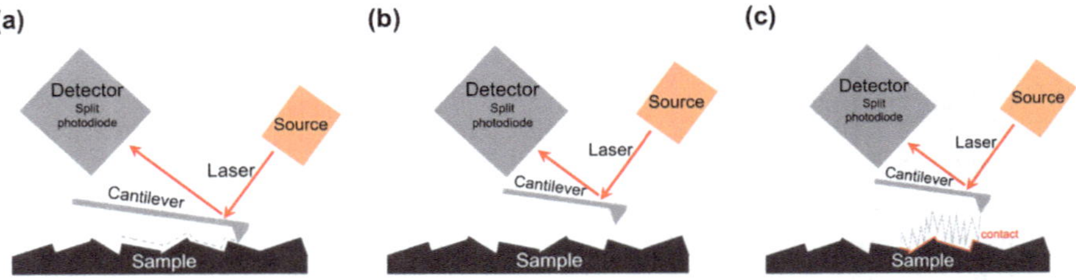

Figure 4.10 Schematic illustration of (a) static or contact AFM mode; (b) non-contact AFM mode; and (c) intermittent or tapping AFM mode, with special focus on the presence of contact during tapping mode and periodic oscillation.

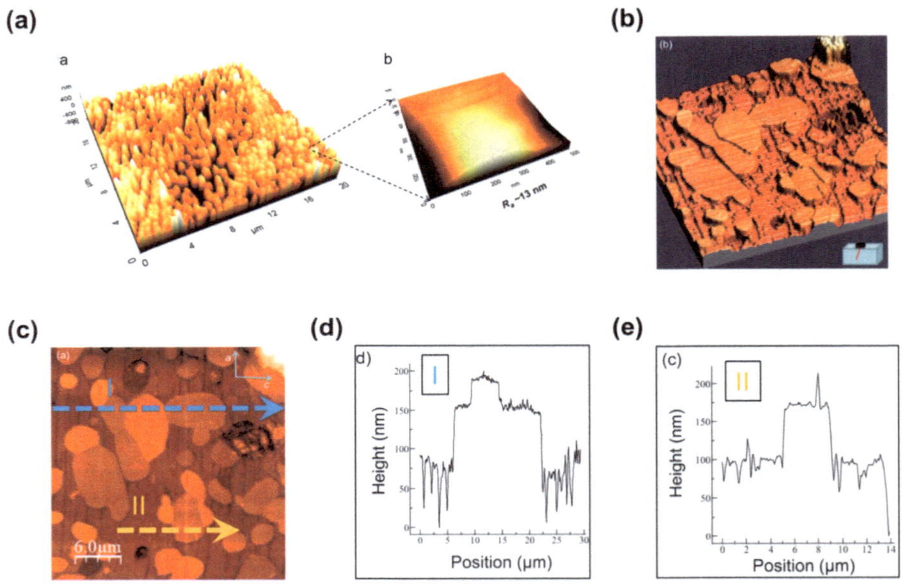

Figure 4.11 (a) AFM imaging of nanoporous Cu foam. Highlight on the smooth surface of a zoomed-in Cu strut with a surface roughness of 13 nm. Reproduced from ref. 47 with permission from Elsevier, Copyright 2014. (b) 3D representation of a Ba–Cu oxide foam. Reproduced from ref. 49 with permission from Springer Nature, Copyright 2014. (c–e) AFM topography image with height profiles of the linear routes I and II. Reproduced from ref. 49 with permission from Springer Nature, Copyright 2014.

and Wang *et al.*[48] used contact AFM mode to quantitatively determine the surface roughness of Cu foams. The AFM images of both Cu materials showed homogeneous morphology and a smooth surface, except near pores, as shown in Figure 4.11(a). Both investigations reported an average surface roughness of 13 and 15.2 nm, respectively (Figure 4.11(b)).[47,48] Considering a linear profile obtained by the I and

II routes in a 2D image (Figure 4.11(c)),[49] it is possible to obtain a 3D map by proceeding with AFM in multiple axial directions, as shown in Figure 4.11(d).

The constant contact between the tip of the cantilever and the sample results in angular variations of its position relative to a laser source. The signal, reflected from the cantilever, is then captured by a detector, providing different readings of the surface roughness of the surface.[47] Moreover, the AFM technique is not able to characterize the entire surface of the metal foam, as the porosity degree and surface topology features are larger than the distance between the surface and the tip of the cantilever, and the latter cannot reach the inner portions of the material, usually limiting the interaction to the outer surface of the material.[49] Furthermore, contact AFM mode is not suitable for samples whose surface cannot be damaged,[50] due to the direct contact between the surface and the cantilever. For that reason, other AFM-based sampling modes have been used for metal foams.

- *Non-contact and tapping AFM modes*

Non-contact AFM characterizes the surface of the material by making use of van der Waals interactions between the tip of the cantilever and the sample surface. Using a small amplitude of oscillation, the cantilever is maintained practically in its equilibrium position without touching the surface of the material, as shown in Figure 4.10(b). Such minimal fluctuations of phase and amplitude of the resonance of the cantilever are converted into measurable data.[46] These Van der Waals interactions are possible with the use of functionalized cantilever tips whose chemical groups interact with those at the surface of the material. CO functionalized tips are widely used because they provide a well-defined atomic termination. By fine-tuning the operational conditions, the composition and geometry of the tip of the cantilever, it is possible to laterally and vertically manipulate the molecules involved, allowing more accurate imaging, as well as giving insights into the properties of the material towards catalysis or other chemical processes.[51]

Metals with highly porous structures are systems where non-contact AFM can be interestingly applied, measuring its roughness and revealing interesting properties/interactions with the surface by means of a functionalized tip. Tapping AFM is a hybrid method of contact and non-contact AFM used to study, for example, samples with a hydrophilic surface to which the tip of the cantilever sticks, compromising the measurement. While maintaining a periodic oscillation, tapping AFM also involves periodic contact with the surface, unlike non-contact AFM, as shown in Figure 4.10(c).[52,53] AFM is a viable method for surface characterization that lacks a competitive presence in metal foams literature most likely due to the common widespread use of other microscopy techniques (such as optical and electron microscopy) that do not include the risk of damaging the samples.[54]

4.1.3.2 Stylus Profilometry

Profilometry is used to characterize the surface roughness of materials. Contact profilometry, also known as stylus profilometry, relies on the application of a set-point force on a stylus in direct contact with the surface of the sample.[55] It allows faster and easier imaging quantifications than AFM techniques at the expense of a lower resolution, approximately an average roughness of 50 to 0.025 µm, according to the International Organization for Standardization (ISO) classification charts (ISO 4287:1997).[56] By going over a region with a very well-defined length, the vibrations of the stylus/surface interaction are measured by the profilometer with a transducer converting the vibrational energy into electrical energy. Then an A/D conversion module is used to convert the electrical signal into digital, thus resulting in the recording of the roughness profile. Although not common, stylus profilometry has been used to characterize the roughness of macroscopic metal foams.[57]

4.1.3.3 Optical Microscopy

There are numerous techniques based on optical microscopy that have been employed to characterize the surface of porous metals. Although an in-depth review of these techniques is out of the scope of the present chapter, it is worth mentioning the most relevant methods alongside their application in the characterization of metallic foams. Besides classical optical microscopy, other techniques of interest based on optical microscopy are confocal laser scanning microscopy and white light interferometry. The combination of these images with specific software for image analysis makes this a simple and popular technique.

- *Optical microscopy*

  Optical microscopy is a widely used and straightforward technique for analysing the morphology of metal foams. This method involves focusing light onto the sample surface, making it an accessible means of observation. Due to the restrictive resolution scope of optical microscopy, the surface observation is, in general, limited to a maximum magnification of 1000×.[44] One example of the application of optical microscopy of metal foams is shown in Figure 4.12(a), where a cobalt foam deposited on stainless steel was observed by means of this technique.
- *Confocal laser scanning microscopy*

  Unlike conventional optical microscopy, confocal laser scanning microscopy (CLSM) is a photonic imaging technique consisting in the fluorescent emission of localized laser-excited points of a given material.[58] The images acquired by CLSM are limited by pixel saturation and photobleaching (the degradation of fluorophores with light), as shown in Figure 4.12(b). When adequate conditions are

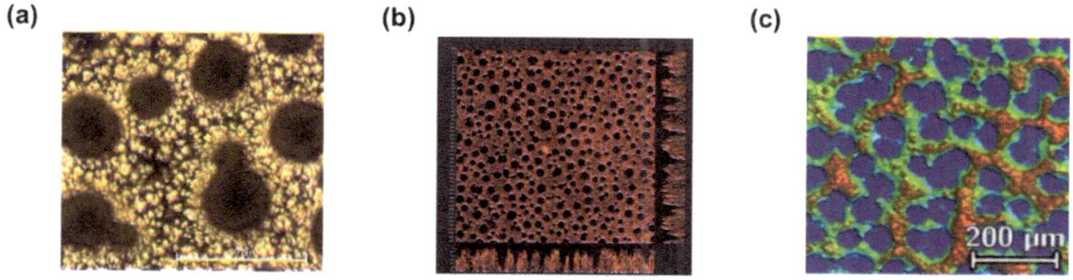

Figure 4.12 (a) Optical microscopy image of a cobalt foam deposited on stainless steel. Source: F. Montemor group. (b) Confocal laser scanning microscopy image of a cobalt foam. Reproduced from ref. 4 with permission from Elsevier, Copyright 2020. (c) White-light interferometric image of a Ag–Cu bimetallic foam deposited on a Cu foil substrate. Reproduced from ref. 66, https://doi.org/10.1002/cctc.202001145, under the terms of the CC BY 4.0 license, https://creativecommons.org/licenses/by/4.0/.

selected, CLSM is capable of producing images with average resolutions of 100 µm.[59] One of the great advantages of using CLSM is its capability of removing out-of-focus light by forcing the image plane through a pinhole, thus focusing on a single irradiated point, which results in a noise-free (absence of out-of-focus light) image by moving the sample or exciting light.[60] In the work of Aryal *et al.*,[61] CLSM was employed for a Cu foam cathode that operated in a bioreactor for hydrogen evolution reactions. The work was able to conclude that hydrogen evolution in the Cu foam was compromised by dead cells of *S. ovata*, detected by CLSM, which were believed to be blocking active sites of the reaction and thus decreasing the activity for hydrogen evolution.

- *White light interferometry*

  White light interferometry is an optical method that splits a white light source into two beams: one into a reference mirror and another to the sample. With equal optical path lengths, the reflections of these beams cause an interference, and, in the case of a rough surface sample, different optical lengths of the measuring beam are detected, creating an interference pattern with the reference beam. By changing the length of the irradiated beams, several interference signals are produced and processed by a charge-coupled device (CCD) that transforms them into a digital signal.[62] Although this method produces quick sub-nanoscale imaging, it requires good reflecting samples, and a lack of angular characteristics pose a disadvantage to study metal foams.[63] White light interferometry can reach resolutions of a few micrometres; however, resolutions at the nanoscale are also reported in the literature.[64,65] As Figure 4.12(c) shows, white light interference can characterize porous metal foams.[66]

4.1.3.4 Electron Microscopy

An overview of two microscopy methods will be given in this section: scanning electron microscopy (SEM) and transmission electron microscopy (TEM). These are based on the interactions between electrons and the surface under study, resulting in images with micro- to nano-scale resolution.

- *Scanning electron microscopy*

  SEM is the most widely-used technique to study the morphology of materials, including metal foams. Electronic microscopy solves the intrinsic limitations of optical microscopy by using electrons to analyse the surface features of materials, instead of light. SEM detects the secondary reflected electrons that scatter upon interacting with the surface of the material, creating a black and white image. The use of electrons allows a much higher resolution, allowing observation down to about 10 nm.[67] SEM imaging is particularly interesting for metal foams, as depicted in Figure 4.13(a–c), because it gives one the ability to measure and estimate the pore dimensions of the 3D architecture, observe the texture of the surface, and detect overall if a certain synthesis pathway resulted in the intended pore distribution and pore size.[4,68] SEM is mostly used to visualize the morphologies after the synthesis of metallic foams is conducted, and is a valuable visual corroboration technique that helps understanding of the type of pores, dendrites and other structures present, in order to evaluate,

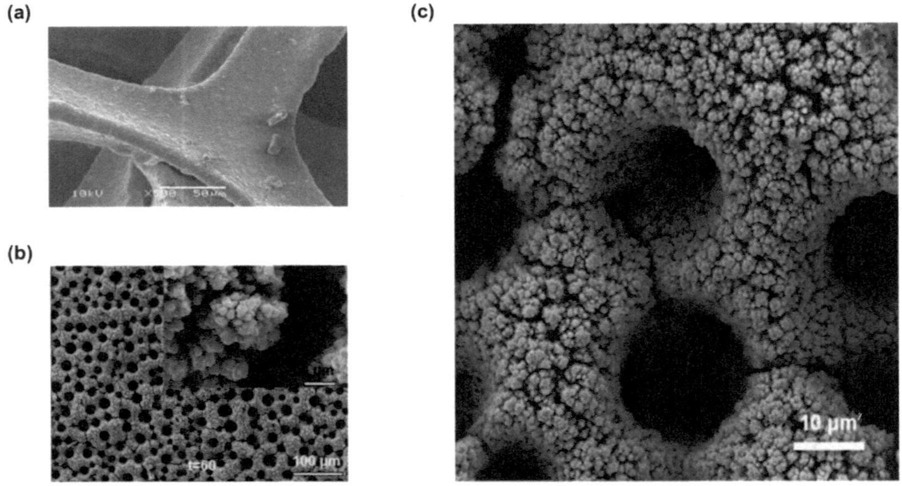

Figure 4.13 SEM images of metal foams: (a) CoNi/rGO on Ni foam. Reproduced from ref. 68 with permission from Elsevier, Copyright 2020. (b) Co foams prepared during 60 s of electrodeposition. Reproduced from ref. 4 with permission from Elsevier, Copyright 2020. (c) Co nanofoam. Reproduced from ref. 4 with permission from Elsevier, Copyright 2020.

adjust and/or completely change the experimental pathway. Combining it with proper modelling and analysis of imaging tools, it is possible to extract relevant quantitative features for better characterization of metal foams.

- *Transmission electron microscopy*

  TEM is another electronic microscopy technique widely used to characterize metal foams. Unlike SEM, this technique transmits electrons through the sample that hit the fluorescent bottom of the microscope creating two-dimensional projections of the inner structure of the material.[68] Therefore, while SEM is used to observe the surface morphology of a material, TEM is applied when a better understanding is required, including structural/crystallographic features. In the work of Li *et al.*,[68] TEM analysis has been used to identify interesting features of Co–Ni based foams, such as metallic wrinkled sheets attributed to the overlap of rGO and Co–Ni nanosheets (Figure 4.14(a)), as well as the presence of nanoholes within the said structure. In another study, Zhao *et al.*[69] studied the presence of rGO on metal foams and concluded that MnO_2 nanoflakes were clearly deposited on the rGO surface *via* the imaging results of TEM (Figure 4.14(b)). In the work of Kim *et al.*,[70] Cu–Ni alloy nanofoams were synthesized, and their structure was found to be polycrystalline based on results obtained by TEM analysis.

Although TEM offers a better understanding of the crystalline structure, impurities and defects, as well as better spatial resolution, the fact that it uses

(a)

(b)

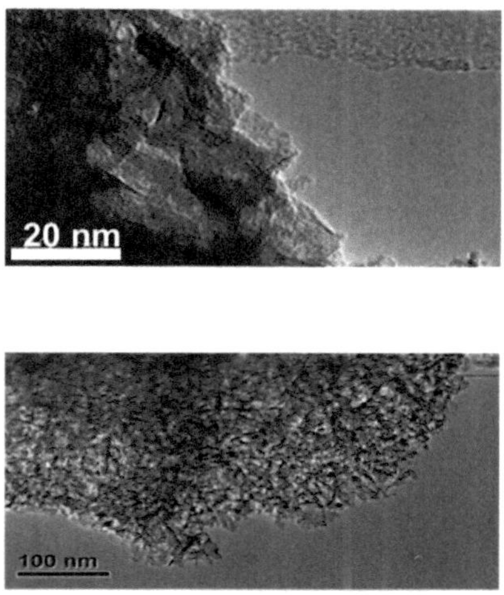

Figure 4.14 TEM images of metal foams: (a) CoNi/rGO on Ni foam. Reproduced from ref. 68 with permission from Elsevier, Copyright 2020. (b) Reduced graphene oxide rGO/MnO_2 foam. Reproduced from ref. 69 with permission from Elsevier, Copyright 2020.

transmitted electrons requires the samples to be very thin (approximately less than 150 nm) in comparison to SEM.[60] Such characteristics limit the application of the method, particularly because of the incompatibility of the TEM setup and the preservation of the sample under certain circumstances.

4.2 Electrical Conductivity

Metallic foams can be obtained from transition metals and have, in general, a great inherent electrical conductivity. Electrical conductivity is a crucial property for systems that depend on charge transfer processes and for capacitive systems. For example, the inherent resistance to the flow of electrical current of a material directly influences the energy and power densities of energy storage devices that use metal foam as electrodes, limiting the scope of the device, that usually require oversizing to reach the target values.[71] Electrical conductivity (σ) is a property that defines the affinity of electric current to flow through a material, while electrical resistivity (ρ) is its opposite. These two parameters (eqn (4.8)) are inversely proportional properties and directly dependent on certain geometrical parameters, according to:

$$\sigma = \frac{1}{\rho} = \frac{l}{R \times A} \tag{4.8}$$

where A is the cross-sectional area in relation to the current flow direction, l is the length that the electric current must flow through, and R is the inherent resistance of the material. The electrical conductivity and resistivity can vary greatly with the composition of the metal foams. The electrical conduction also depends on the extrinsic properties of the metals, particularly, their morphology. Foam structural distortions and porosity have been reported to have a great influence in the real electrical conductivity of open- and closed-cell foams.[72]

Resistive impurities (impurities with lower electrical conductivity than the base material) and morphological defects will also negatively influence conductivity. It is worth noting that pores are electrically insulating. By reducing the geometrical volume through which electrons can diffuse, the overall electrical conductivity will be lower. Cuevas *et al.*[72] demonstrated this phenomenon by showing the clear decrease of conductivity of a diverse selection of metal foams as their porosity increases (Figure 4.15). In the work of Clyne *et al.*,[73] a comprehensive study on the fundamentals of thermal and electrical conductivity of metal matrix composites concludes that the conductivities of matrix constituents and the geometry of the structure are sufficient to develop easily understood analytical models that enable the determination of accurate overall composite electrical and thermal conductivities. The work further states that the estimated conductivity values can significantly differ from reality due to differences in the microstructure, such as structural defects. Nevertheless, there is a clear confidence level that the conductivity of 3D structured composites can be effectively determined, especially for pure metallic foams.[73]

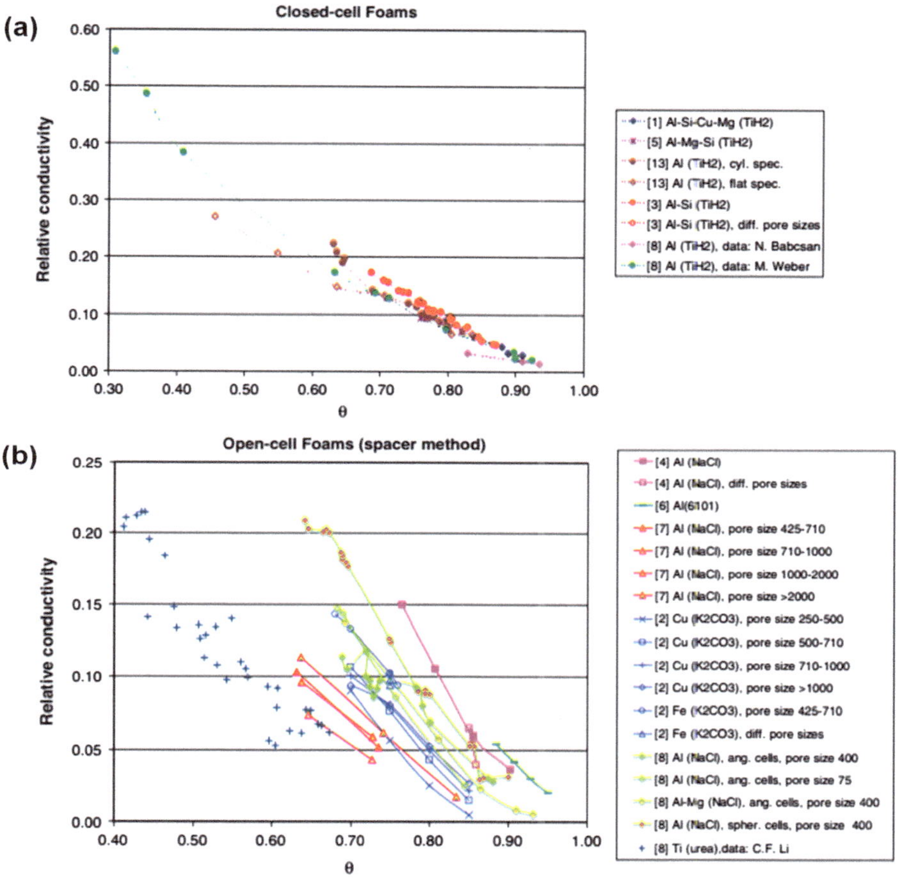

Figure 4.15 Electrical conductivity as a function of porosity of (a) closed-cell and (b) open-cell metal foams. Reproduced from ref. 72 with permission from Springer Nature, Copyright 2008.

To measure the electrical conductivity of metal foams, the three most referenced methods in the literature are the four-point-probe method (4PP), the Eddy current testing (ECT) method and electrochemical impedance spectroscopy (EIS), which is based on the conductivity working principle of ECT.[72,74–77]

- *Four-point-probe method*

  The four-point-probe method (Figure 4.16(a)) relies on four probes equally spaced that are connected to the surface of the material side by side. The two outer probes act as current source and sensing, respectively, and the two inner probes measure the resulting voltage drop between them. The separation of current and voltage allows the

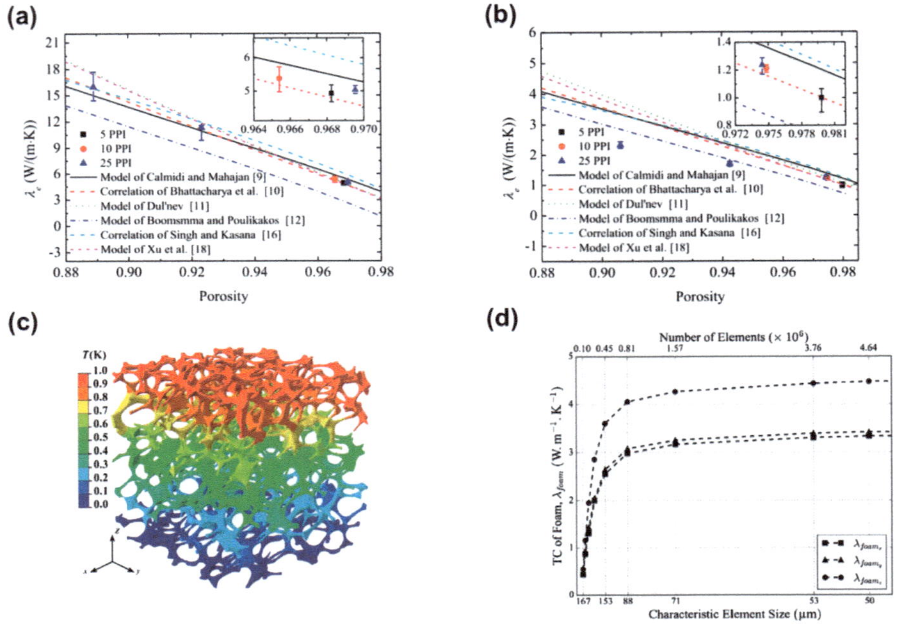

Figure 4.16 (a and b) Dependence of thermal conductivity on the porosity for Cu and Ni foams, respectively. Reproduced from ref. 83 with permission from Elsevier, Copyright 2014. (c) 3D steady-state model of an Al foam temperature profile based on a Fourier model of heat transfer and (d) the calculated thermal conductivity across the spatial axis of the Al foam. Reproduced from ref. 84 with permission from Elsevier, Copyright 2018.

calculation of R with which the electrical conductivity or resistivity are then calculated with eqn (4.8). In the case of metal foams, the relative conductivity of metal foams decreases as porosity increases, as expected and concluded in the work of Cuevas *et al.* and showcased in Figure 4.15(a).[72] Dharmasena *et al.*[74] also demonstrated this relationship *via* the 4PP method for Al foams and determined that the electrical conductivity varies linearly with porosity. The same method can be used to measure the thickness of the material when its resistance is known.

- *Eddy current testing method*

  The Eddy current testing method is a technique that works on the principles of Faraday's law of electromagnetic induction. In ECT, a magnetic field is generated around a coil (probe) when a current is applied.[74] When brought close to a surface, such as the surface of a metal foam, the magnetic field induces currents to the surface and the near depths of the material – known as the Eddy currents[78] – which are detected by the test coil. In the work of Babcsán *et al.*,[75] ECT was concluded to be a useful and cost-effective tool to determine the electrical conductivity of different Al foams, which correlated a conductivity increase with the relative density of the foams.

- *Electrochemical impedance spectroscopy*

  Electrochemical impedance spectroscopy is widely used to study the AC resistance, *i.e.*, impedance, of electrochemical systems. It can also be used to indirectly determine the electrical conductivity of a material by using known mathematical relationships between conductivity and resistance. The latter can be calculated from EIS measurements by fitting the experimental impedance data with a suitable equivalent circuit model. In the work of Ma *et al.*,[76] the electrical conductivity was first measured *via* ECT to demonstrate that the morphological traits of Cu and Fe foams (*e.g.*, relative density, porosity) considerably influence the eddy currents, ultimately interfering with the impedance results.

EIS, similarly to CV, can also be used to calculate double layer contributions, correlated to the electrochemical surface area, by comparing the spectra obtained for a flat polished electrode to that of the foam material (it is worth noting that this must be performed in an electrolyte that does not introduce relevant faradic contributions).[29] By fitting these spectra with an equivalent circuit model, it is possible to determine the double layer capacitance and therefore the area of the electrode contributing to it. This technique is quite reliable, and the averaged response of the working electrode can be explored to estimate the surface area of the foams. More details on the fundamentals regarding EIS and its applications can be found elsewhere.[79,80]

4.3 Thermal Conductivity

Analogous to electrical conductivity, thermal conductivity is an inherent property of metals that is highly influenced by the material's composition and morphology. Thermal conductivity (TC) is a measure of the ability of the material to conduct energy in the form of heat. Considering their dependence on composition, mixtures of metals may increase or decrease the overall thermal conductivity of the overall composite. Temperature, phase changes and crystalline anisotropy are other key factors that affect thermal conductivity, as extensively discussed by David W. Hahn in ref. 81.

Porous materials, such as metal foams, are attracting a lot of attention for heat exchanging purposes. This has driven researchers to better understand the heat transfer phenomena on 3D structures, which can be seen in the literature.[82–85] In the work of Xiao *et al.*,[83] the thermal conductivity of metal foams decreased with its porosity (Figure 4.16(a) and (b)) as further geometrical pore parameters (pore size and distribution) had a negligible effect on this property.

The estimation of thermal conductivity relies on the application of Fourier's model for heat transfer. Thermal conductivity is mathematically calculated from a temperature distribution (experimental or simulated).[84] This observation can be found in the work Amani *et al.*[84] where a mathematical model was applied to determine the thermal conductivity within the scope of the 3D modelling of a metal foam (Figure 4.16(c) and (d)).

4.4 Mechanical Properties

Considering their inherent 3D architectures, metal foams have interesting mechanical properties that define their behaviour upon being submitted to mechanical stress. Several properties establish the mechanical stability of a given material: strength, stiffness, elasticity, plasticity, ductility, brittleness, malleability, toughness, hardness and fatigue, among others; properties reviewed by Murugan *et al.*[86] An understanding of mechanical properties is generally established by experimental stress–strain curves. These are obtained by measuring the variation of stress and strain of the material as an external force is applied. The stress, σ, is the restoring, or deforming, pressure (Pa) exercised in the material, while strain, ε, is the length of the material divided by its original length (%).

Jiang *et al.*[87] studied the mechanical behaviour of Ni–Ag foams, showcasing three clear regions in their stress–strain curves (compression test): elastic, densification and collapse. Firstly, a rapidly increasing and practically linear region, which ends at approximately 0.05% of strain, defines the elastic region, the stress–strain point up to which the metal foam can resume its initial geometrical structure without deformities or any kind of changes. Secondly, a clear plateau region defines the plastic region, which is the stress–strain region where the metal foam will no longer return to its initial form and its structure will continuously deform at constant stress until the material collapses, evidenced by the sudden stress peak at the end of the curve. Jiang *et al.*[87] further demonstrated that Ni–Ag foams of lower density have a lower stress plateau but a higher densification strain, as observed for the Ni–Ag foam of 15.8 mg cm^{-1} (Figure 4.17(a)). In another work, Martínez *et al.*[88]

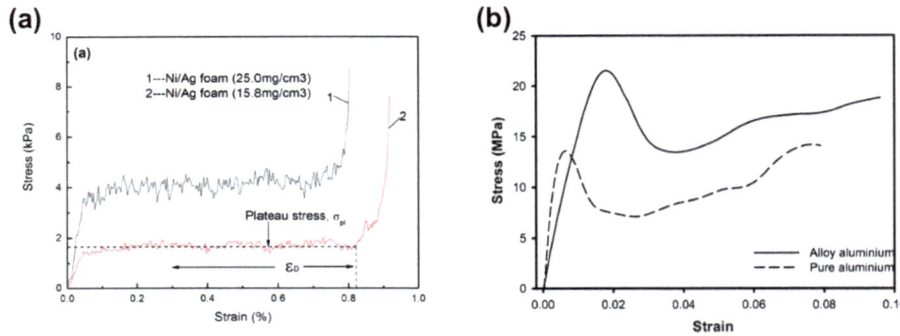

Figure 4.17 Stress–strain curves of (a) Ni–Ag foams with different densities. Reproduced from ref. 87, https://doi.org/10.1038/srep13825, under the terms of the CC BY 4.0 License, https://creativecommons.org/licenses/by/4.0/. (b) Pure Al foam and alloy-Al foam. Reproduced from ref. 88 with permission from Springer Nature, Copyright 2016.

conducted compression tests of pure Al and alloyed Al foams and demonstrated that alloyed Al foams (alloying elements: Ca, Ti, Fe, Mn) are more stress–strain resistant to the pure Al foam, as observed in Figure 4.17(b).

Stress–strain curves can be obtained by a wide array of mechanical tests: indentation, elongation, bending and compression tests, represented in Figure 4.18(a–d), respectively.

Indentation testing can be used to measure a material's hardness facing a controlled deformation. It is a versatile testing method that can be applied to macro- or nano-sized structures, which, in the latter case, are commonly referred as nanoindentation. The working principle is simple: an indenter (the material that will cause the deformation, usually diamond-based) will act as a force against the sample. The resulting indentation (defined by the depth and length – d and Φ in Figure 4.18(a)) is measured as a function of the force applied, creating a stress–strain curve.[87–98] In the work of Linul *et al.*[98] composite panels with Al foam core were studied *via* indentation testing and it was concluded that rigid-supported panels better absorbed the energy than fully fixed panels, suggesting an optimal sandwich-type of panel/foam structure for energy absorption.

Elongation tests consist in the study of the ductility of the material by application of stretching forces in opposite directions (Figure 4.18(b)), deforming the material until a state of rupture is achieved, thus interrupting the stress–strain plot.[92–94] Liu *et al.*[92] showed in their work that the relative elongation of Ni foams can be improved *via* chemical plating.

Bending tests are used in the study of the ductility and brittleness of materials and consist in pushing down a material held by one to four points of fixed

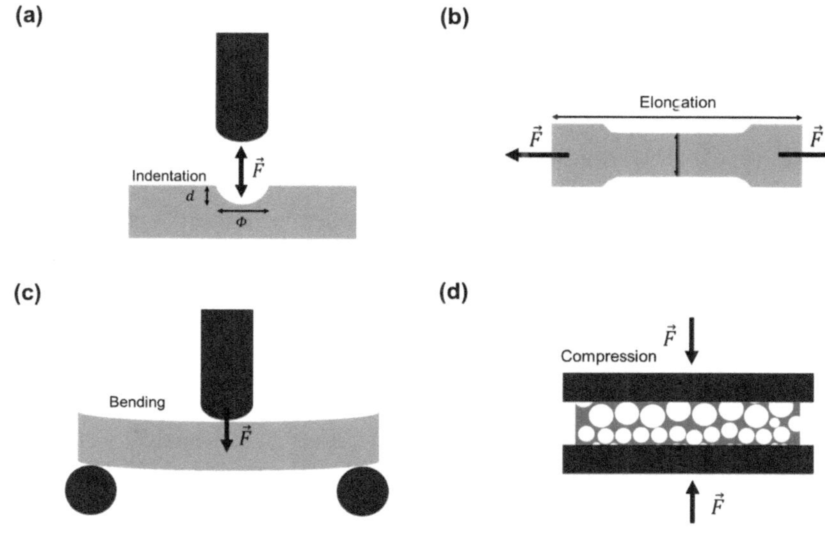

Figure 4.18 Mechanical characterization methods: (a) indentation test, (b) elongation/tensile test, (c) 3-point bending test and (d) compression test.

structure, resulting in a bend, and/or a fractural event (Figure 4.18(c)).[95–97] The study of Kashef *et al.*[97] showed that a mathematical bending model applied to Al alloy foams is also applicable for Ti foams, an important metal for biomedical applications. In the same work, a compression test of Ti foam was also shown. Compression tests are based in the opposite process of elongation and are used to study a material's behaviour facing a crushing load (Figure 4.18(d)).[87,88,98] The increasing porosity of metal foams results in lower mechanical strength, evidence further supported in the literature as a possible consequence of the coalescence of pores and voids, thus crippling the mechanical integrity.[99–101]

Another important property obtained from stress–strain curves is the Young's modulus. The Young's modulus, E, is the proportionality constant of the elasticity domain, the initial linear region commonly observed in stress–strain curves, and is calculated by dividing stress over strain, as described by eqn (4.9):

$$E = \frac{\sigma}{\varepsilon} \tag{4.9}$$

A high Young's modulus effectively characterizes the material as resilient to stress. The Young's modulus of metal foams, and thus, its elasticity, often decreases as the porosity of the foams increases. For example, in the work of Kovacik *et al.*,[102] the Young's modulus of Al–Si based metal foams considerably increased with an increase of the material's density (and, logically, decrease of porosity), going from 4.93 GPa for ~84% of porosity to a much higher 70 GPa for 0%.

4.5 Final Thoughts on the Characterization of Metallic Foams

As exemplified in this chapter, metal foams can be characterized by multiple methodologies. However, it is worth noting that all the techniques reviewed in this chapter were somewhat related to one main characteristic of metal foams: their surface area. Metal foams can be studied *via* adsorption or electrochemical reactions. Methods such as BET or hydrogen adsorption rely on active sites in the foam's surface, whose interaction with nitrogen or hydrogen under the appropriate conditions allow an accurate understanding of the pore size, pore distribution and the SSA of the metal foam. Imaging techniques such as optical microscopy, SEM or TEM allow the visual characterization of metal foams with great resolution and level of detail, enabling one to make conclusions about the morphology, and possibly enabling the observation of interesting nanostructures, which may lead to the development of novel functionalization pathways to further optimize the material's SSA.

Key properties considered in the application of 3D metal foams in the fields of energy storage and heat transfer can be explored by studying their electrical and thermal conductivities through methods that use electromagnetics and complex mathematical models to calculate and characterize these properties across 3D structures. And finally, the mechanical properties of metal foams that can be studied by diverse stress–strain tests, such as indentation, elongation, bending and compression, characterize the elasticity, plasticity, resilience, and maximum stress levels that metal foams can support. It is interesting to observe the direct correlation between porosity and their electrical, thermal or mechanical performance; and the importance of considering the trade-off between porosity/SSA and other key properties when considering metal foams for specific applications.

References

1. H. Sun, J. Zhu, D. Baumann, L. Peng, Y. Xu, I. Shakir, Y. Huang and X. Duan, *Nat. Rev. Mater.*, 2018, **4**, 45–60.
2. R. Dubey and V. Guruviah, *Ionics*, 2019, **25**, 1419–1445.
3. R. A. Fisher, M. R. Watt and W. Jud Ready, *ECS J. Solid State Sci. Technol.*, 2013, **2**, M3170–M3177.
4. P. Arévalo-Cid, A. Adán-Más, T. M. Silva, J. A. Rodrigues, E. Maçôas, M. F. Vaz and M. F. Montemor, *Mater. Charact.*, 2020, **169**, 110598.
5. D. Whitehouse, *Surfaces and Their Measuremente*, 1st edn, 2004.
6. P. M. V. Raja and A. R. Barron, *LibreTexts*, 2021.
7. M. Thommes, K. Kaneko, A. V. Neimark, J. P. Olivier, F. Rodriguez-Reinoso, J. Rouquerol and K. S. W. Sing, *Pure Appl. Chem.*, 2015, **87**, 1051–1069.
8. K. S. W. Sing, *Colloids Surf.*, 2001, 187–188.
9. A. Datar, Y. G. Chung and L. C. Lin, *J. Phys. Chem. Lett.*, 2020, **11**, 5412–5417.
10. J. Lee and Y. C. Choi, *Materials*, 2020, **13**, 4038.
11. D. Gao, G. Yang, Z. Zhu, J. Zhang, Z. Yang, Z. Zhang and D. Xue, *J. Mater. Chem.*, 2012, **22**, 9462–9465.
12. J. R. Rajabathar, H. A. Al-lohedan, P. Arunachalam, Z. A. Issa, M. K. Gnanamani, J. N. Appaturi, S. N. Ibrahim and W. Mohammed Dahan, *J. Alloys Compd.*, 2021, **850**, 156346.
13. J. M. Haynes, *Mater. Constr.*, 1973, **6**, 209–213.
14. O. Maalal, M. Prat and D. Lasseux, *Phys. Fluids*, 2021, **33**, 027103.
15. R. Bardestani, G. S. Patience and S. Kaliaguine, *Can. J. Chem. Eng.*, 2019, **97**, 2781–2791.
16. Y. Li, Z. Zhou, H. Xu, X. Liao, G. Cheng, M. Sun and L. Yu, *J. Mater. Sci.: Mater. Electron.*, 2019, **30**, 18424–18431.
17. S. Palchoudhury, M. Baalousha and J. R. Lead, in *Characterization of Nanomaterials in Complex Environmental and Biological Media*, ed. M. Baalousha and J. R. Lead, Frontiers of Nanoscience, Elsevier, 2015, vol. 8, pp. 153–181.
18. C. Nguyen and D. D. Do, *Carbon*, 2001, **39**, 1327–1336.
19. R. J. Dombrowski, C. M. Lastoskie and D. R. Hyduke, *Colloids Surf., A*, 2001, **187**, 23–39.

20. T. Bernet, M. M. Piñeiro, F. Plantier and C. Miqueu, *Mol. Phys.*, 2020, **118**, e1767308.
21. A. L. Kolesnikov, Y. A. Budkov and G. Y. Gor, *J. Phys. Chem. C*, 2020, **124**, 20046–20054.
22. L. Zhang, Y. Xiong, Y. Li, M. Wei, W. Jiang, R. Lei and Z. Wu, *Fuel*, 2017, **204**, 1–11.
23. J. M. D. Rodríguez, J. A. H. Melián and J. P. Peña, *J. Chem. Educ.*, 2000, **77**, 1195–1197.
24. P. Simon and Y. Gogotsi, *Nat. Mater.*, 2008, **7**, 845–854.
25. P. Connor, J. Schuch, B. Kaiser and W. Jaegermann, *Z. Phys. Chem.*, 2020, **234**, 979–994.
26. K. I. Siwek, S. Eugénio, D. M. F. Santos, M. T. Silva and M. F. Montemor, *Int. J. Hydrogen Energy*, 2019, **44**, 1701–1709.
27. K. K. Diao, Z. Xiao and Y. Y. Zhao, *Mater. Chem. Phys.*, 2015, **162**, 571–579.
28. F. Wang, T. A. Shifa, X. Zhan, Y. Huang, K. Liu, Z. Cheng, C. Jiang and J. He, *Nanoscale*, 2015, **7**, 19764–19788.
29. S. Trasatti and O. A. Petrii, *J. Electroanal. Chem.*, 1992, **327**, 353–376.
30. J. Ekspong, E. Gracia-Espino and T. Wågberg, *J. Phys. Chem. C*, 2020, **124**, 20911–20921.
31. M. Łukaszewski, *Int. J. Electrochem. Sci.*, 2016, 4442–4469.
32. C. Schlaup and S. Horch, *Phys. Chem. Chem. Phys.*, 2013, **15**, 19659–19664.
33. G. I. Golodets, in *Studies in Surface Science and Catalysis*, Elsevier, 1983, vol. 15, pp. 280–311.
34. S. Rudi, C. Cui, L. Gan and P. Strasser, *Electrocatalysis*, 2014, **5**, 408–418.
35. T. Binninger, E. Fabbri, R. Kötz and T. J. Schmidt, *J. Electrochem. Soc.*, 2014, **161**, H121–H128.
36. X. Ren, Q. Lv, L. Liu, B. Liu, Y. Wang, A. Liu and G. Wu, *Sustainable Energy Fuels*, 2020, **4**, 15–30.
37. C. Zhou, L. Tao, F. Yang, B. Wang, X. Wan, Y. Jin, H. Yu and Y. Yang, *Curr. Opin. Chem. Eng.*, 2019, **26**, 88–95.
38. W. Plieth, *Electrochem. Mater. Sci.*, 2008, **4**, 101–142.
39. B. Madry, M. Nowicki and K. Wandelt, Sulfate Adsorption on Copper: From upd to Bulk Copper on Au(111), in *Encyclopedia of Interfacial Chemistry: Surface Science and Electrochemistry*, 2018, vol. 5, pp. 281–291.
40. N. Mayet, K. Servat, K. B. Kokoh and T. W. Napporn, *Surfaces*, 2019, **2**, 257–276.
41. M. Li, Q. Ma, W. Zi, X. Liu, X. Zhu and S. F. Liu, *Sci. Adv.*, 2015, **1**, e1400268.
42. J. Xu, R. Li, C.-Q. Xu, R. Zeng, Z. Jiang, B. Mei, J. Li, D. Meng and J. Chen, *Appl. Catal., B*, 2021, **289**, 120028.
43. M. Chatenet, Y. Soldo-Olivier, E. Chaînet and R. Faure, *Electrochim. Acta*, 2007, **53**, 369–376.
44. S. M. Alia, K. E. Hurst, S. S. Kocha and B. S. Pivovar, *J. Electrochem. Soc.*, 2016, **163**, F3051–F3056.
45. R. Asmatulu and W. S. Khan, Characterization of electrospun nanofibers, in *Synthesis and Applications of Electrospun Nanofibers*, Elsevier, 2019, pp. 257–281.

46. P. Eaton and P. West, *Atomic Force Microscopy*, Oxford University Press Inc., New York, United States, 2010.
47. H. Park, C. Ahn, H. Jo, M. Choi, D. S. Kim, D. K. Kim, S. Jeon and H. Choe, *Mater. Lett.*, 2014, **129**, 174–177.
48. Z. Wang, O. Yaegashi, H. Sakaue, T. Takahagi and S. Shingubara, *J. Electrochem. Soc.*, 2004, **151**, C781–C785.
49. M. R. Koblischka, A. Koblischka-Veneva, E. S. Reddy and G. J. Schmitz, *J. Adv. Ceram.*, 2014, **3**, 317–325.
50. J. V. Lauritsen and M. Reichling, *J. Phys.: Condens.Matter*, 2010, **22**, 263001.
51. L. Gross, F. Mohn, N. Moll, P. Liljeroth and G. Meyer, *Science*, 2009, **325**, 1110–1114.
52. Y. Martin, C. C. Williams and H. K. Wickramasinghe, *J. Appl. Phys.*, 1987, **61**, 4723–4729.
53. Q. Zhong, D. Inniss, K. Kjoller and V. B. Elings, *Surf. Sci. Lett.*, 1993, **290**, L688–L692.
54. É. Maire, J. Adrien and C. Petit, *C. R. Phys.*, 2014, **15**, 674–682.
55. T. V. Vorburger, H. G. Rhee, T. B. Renegar, J. F. Song and A. Zheng, *Int. J. Adv. Des. Manuf. Technol.*, 2007, **33**, 110–118.
56. I. O. F. Standardization, ISO 4287:1997 Geometrical Product Specifications (GPS) — Surface texture: Profile method — Terms, definitions and surface texture parameters, 2021.
57. M. Garcia-Avila and A. Rabiei, *Metals*, 2015, **5**, 822–835.
58. A. Canette and R. Briandet, in *Encyclopedia of Food Microbiology*, Elsevier, 2nd edn., 2014, pp. 676–683.
59. A. Barbeau, J. L. Swift, A. G. Godin, Y. De Koninck, P. W. Wiseman and J. M. Beaulieu, *Methods Cell Biol.*, 2013, **117**, 1–19.
60. C. M. St Croix, S. H. Shand and S. C. Watkins, *Biotechniques*, 2005, **39**, S2–S5.
61. N. Aryal, PhD, Technical University of Denmark, Denmark, 2017.
62. E. Novak, D.-S. Wan, P. Unruh and J. Schmit, *J. Phys.: Condens. Matter*, 2010, **22**, 263001.
63. Q. J. Wang and Y.-W. Chung, *Encyclopedia of Tribology*, 2013.
64. P. Lehmann, S. Tereschenko and W. Xie, *Surf. Topogr.: Metrol. Prop.*, 2016, **4**, 024004.
65. F. Wang, L. Liu, P. Yu, Z. Liu, H. Yu, Y. Wang and W. J. Li, *Sci. Rep.*, 2016, **6**, 24703.
66. S. Vesztergom, A. Dutta, M. Rahaman, K. Kiran, I. Zelocualtecatl Montiel and P. Broekmann, *ChemCatChem*, 2020, **13**, 1039–1058.
67. W. Zhou, R. Apkarian, Z. L. Wang and D. Joy, in *Scanning Microscopy for Nanotechnology*, ed. W. Zhou, and Z. L. Wang, Springer, New York, NY, 2006, pp. 1–40.
68. B. Li, C. Song, J. yin, J. Yan, K. Ye, K. Cheng, K. Zhu, D. Cao and G. Wang, *Int. J. Hydrogen Energy*, 2020, **45**, 10569–10579.
69. Z. Zhao, T. Shen, Z. Liu, Q. Zhong and Y. Qin, *J. Alloys Compd.*, 2020, 812.

70. C.-E. Kim, R. M. Rahimi and D. F. Bahr, *MRS Commun.*, 2020, **10**, 286–291.
71. H. T. Sun, J. Zhu, D. Baumann, L. L. Peng, Y. X. Xu, I. Shakir, Y. Huang and X. F. Duan, *Nat. Rev. Mater.*, 2019, **4**, 45–60.
72. F. G. Cuevas, J. M. Montes, J. Cintas and P. Urban, *J. Porous Mater.*, 2008, **16**, 675–681.
73. T. W. Clyne, in *Comprehensive Composite Materials II*, ed. P. W. R. Beaumont and C. H. Zweben, Elsevier, 2018, ch. 7, vol. 4, pp. 188–212.
74. K. P. Dharmasena and H. N. G. Wadley, *J. Mater. Res.*, 2002, **17**(03), 625–631.
75. N. Babcsán, I. Mészáros and N. Hegman, *Mater. Sci. Eng. Technol.*, 2003, **34**, 391–394.
76. X. Ma, A. J. Peyton and Y. Y. Zhao, *NDT&E Int.*, 2006, **39**, 562–568.
77. X. L. Huang, G. H. Wu, Z. Lv, Q. Zhang, P. C. Kang and J. F. Leng, *J. Alloys Compd.*, 2009, **479**, 898–901.
78. P. C. R. Bhagi, *J. Nondestr. Test. Eval.*, 2011, **10**, 7–16.
79. M. E. Orazem and B. Tribollet, *Electrochemical Impedance Spectroscopy*, John Wiley & Sons, Inc., Hoboken, New Jersey, 2008.
80. A. J. Bard and L. R. Faulkner, *Electrochemical Methods: Fundamentals and Applications*, John Wiley & Sons, Inc., United States of America, 2000.
81. D. W. Hahn and M. N. Özisik, *Heat Conduction*, Wiley, 3rd edn, 2012.
82. T. W. Clyne, I. O. Golosnoy, J. C. Tan and A. E. Markaki, *Philos. Trans. R. Soc., A*, 2006, **364**, 125–146.
83. X. Xiao, P. Zhang and M. Li, *Int. J. Therm. Sci.*, 2014, **81**, 94–105.
84. Y. Amani, A. Takahashi, P. Chantrenne, S. Maruyama, S. Dancette and E. Maire, *Int. J. Heat Mass Transfer*, 2018, **122**, 1–10.
85. T. Fend, J. Sauerhering, A. K. S. do Couto, R. Pitz-Paal and S. Angel, presented in part at the 17th European Conference on Thermophysical Properties, 2005.
86. S. S. Murugan, *Int. J. Mod. Stud. Mech. Eng.*, 2020, **6**, 28–38.
87. B. Jiang, C. He, N. Zhao, P. Nash, C. Shi and Z. Wang, *Sci. Rep.*, 2015, **5**, 13825.
88. V. Prieto Martínez, J. Torres Torres and A. Flores Valdés, *J. Porous Mater.*, 2016, **24**, 707–712.
89. A. Nayebi and A. Surmiri, *Metall. Res. Technol.*, 2018, **115**, 405.
90. J. Němeček and V. Kralik, *Key Eng. Mater.*, 2015, **662**, 59–62.
91. D. Tzetzis, K. Tsongas and G. Mansour, *Mater. Res.*, 2017, **20**, 1571–1578.
92. P. S. Liu, *Mater. Sci. Technol.*, 2013, **20**, 669–672.
93. C. Motz and R. Pippan, *Acta Mater.*, 2001, **49**, 2463–2470.
94. H. Von Hagen and W. Bleck, *MRS Online Proc. Libr.*, 1998, **59**, 64.
95. J. A. Brown, L. J. Vendra and A. Rabiei, *Metall. Mater. Trans. A*, 2010, **41**, 2784–2793.
96. A. Timari, M. Berczeli, J. Liska and Z. Weltsch, *IOP Conf. Ser.: Mater. Sci. Eng.*, 2018, **448**, 012067.
97. S. Kashef, S. A. Asgari, P. D. Hodgson and W. Y. Yan, *Adv. Mater. Res.*, 2008, **32**, 237–240.

98. J. Kováčik, L. Marsavina, A. Adamčíková, F. Simančík, R. Florek, M. Nosko, P. Tobolka, P. Minár, N. Mináriková, J. Jerz and E. Linul, *Key Eng. Mater.*, 2014, **601**, 237–241.
99. X. C. Xia, X. W. Chen, Z. Zhang, X. Chen, W. M. Zhao, B. Liao and B. Hur, *J. Magnesium Alloys*, 2013, **1**, 330–335.
100. S. N. Aqida, M. I. Ghazali and J. Hashim, *J. Teknol.*, 2004, **40**(A), 17–32.
101. G. W. Mugica, D. O. Tovio, J. C. Cuyas and A. C. González, *Mater. Res.*, 2004, **7**, 221–229.
102. J. Kovacik, L. Marsavina and E. Linul, *Materials*, 2018, **11**, 1904.

CHAPTER 5

Supercapacitor Applications

5.1 Introduction

The advantages of using metallic foams for the fabrication of supercapacitors include the possibility to enlarge the volumetric mass loading and the active surface area and the possibility to create pore distributions that enable faster electrochemical processes. Usually, two-dimensional current collectors, for instance, aluminium and copper, are used in the preparation of the electrode. However, in the preparation of the high mass loading (>5 mg cm^{-2}) of active material demanded for energy applications, relatively large amounts of binder are required to ensure the attachment of the active material onto the surface. Such excess of binder negatively affects the electrochemical performance of the active material, for instance, by decreasing the movement of ions into and out of the structure leading to an increase in the resistance. Moreover, during cycling, slow deterioration of conventional electrodes occurs, owing to low mechanical strength that leads to low adherence and separation of the active material from the substrate, as shown in Figure 5.1(a).[1] Therefore, building self-supported electrode materials on top of 3D foams provides good adhesion, and enhanced cyclic stability and rate capability in electrochemical energy storage devices such as supercapacitors and batteries, as depicted in Figure 5.1(b). Moreover, the easy functionalization of metallic and carbon nanofoams with active materials has been considered as a step forward in the rational design of supercapacitor electrodes tailored for application requirements. Nonetheless, a current challenge is maintaining the energy and power capabilities of less porous materials in commercial devices that ought to perform at very high currents.

Transition Metal-based Nanofoams for Electrochemical Systems: Manufacturing, Characterization and Applications
By Maria de Fátima Montemor, Diana M. Fernandes, Alberto Adan-Mas, Ana Catarina Alves, Gabriel Garcia Carvalho and Inês S. Marques

Published by the Royal Society of Chemistry, www.rsc.org

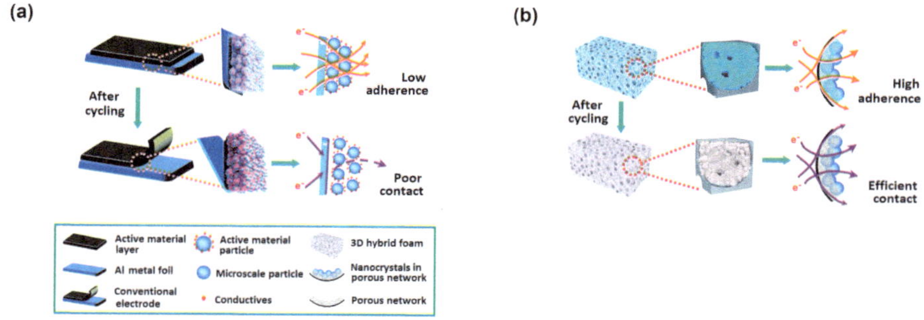

Figure 5.1 Schematics of the preparation of conventional (a) *vs.* foam-based (b) supported cathode materials for Na-ion batteries. Reproduced from ref. 1 with permission from the Royal Society of Chemistry.

Supercapacitors (SCs) have become essential systems for driving the energetic transition towards sustainable energy management, as they offer high-power capabilities and long-term cycling stability. The interesting balance between high-energy and high-power electrochemical response of SCs relies on fast and highly reversible surface processes, which also result in stable specific capacitance and long cycle life.[2–4] Traditional supercapacitors are made of electrodes that use carbon as the active material. Since the specific capacitance of carbon is modest, and in order to increase the energy density, an organic electrolyte is used. This allows one to increase the voltage window of the device and, consequently, the energy density. However, many organic electrolytes are unstable, often toxic and not environmentally friendly. They also pose additional barriers to the sustainable recycling of supercapacitors that nowadays are considered electronic waste. Thus, there has been a lot of pressure to develop alternative electrolytes and materials, combining both high specific power and high specific energy.

Aqueous redox supercapacitors, which utilize eco-friendly aqueous electrolytes with high ionic conductivity, are becoming a promising solution for energy storage applications due to their ability to store and release large amounts of charge quickly and sustain long cycling life. Additionally, these tailored devices can be more eco-friendly and also more sustainable along their entire life cycle. The key to these devices' performance lies in the electrode materials and electrolytes, and their correct assembly. Redox-active metal compounds, especially oxides, hydroxides and sulfides, are very interesting candidates for preparing electrodes for supercapacitors because they can store additional charge through fast and reversible redox reactions. Depending on the type of metal compound, these reactions can be more superficial or involve intercalation of relevant ions such as the proton in the electrolyte. Oxide materials, as is the case of ruthenium oxide or manganese oxides, have been considered very promising materials due to their pseudocapacitive electrochemical response, quasi reversible and fast faradic processes. These processes contribute to increasing the specific capacitance and therefore the energy that the device can store. The electrochemical performance and device metrics can be further improved when the electrode materials possess high surface area. This means that more redox active sites can be exposed to the electrode contributing to the charge storage process. For that, 3D assemblies bring important competitive features. Both carbon and metal nanofoams are immensely attractive as electroactive materials for the construction of supercapacitor electrodes. Despite the different storage mechanisms of charge storage, carbon and metal foams bring important competitive advantages in the assembly of higher performance supercapacitors mainly due to their increased surface area.

5.2 Fundamental Principles of Supercapacitors

Supercapacitors store energy through electrochemical processes, and, based on their charge storage mechanism, they are traditionally divided into electrochemical double-layer capacitors (EDLCs) and pseudocapacitors. Additionally, supercapacitors can be classified into two main categories: symmetrical and asymmetrical. Symmetrical capacitors use the same electrode material for both electrodes, as is the case of the conventional EDLCs and have been commonly used and marketed. Asymmetrical capacitors are commercially available, but not to such a large extent and the main feature is that they use different materials for the cathode and the anode. It is important to note that the electrolytes used in these capacitors can be either organic, such as acetonitrile or propylene carbonate, or aqueous, such as potassium hydroxide or sulfuric acid. However, in commercial devices, mostly the organic ones are used in order to enlarge the voltage window and enhance the specific energy. Nowadays, asymmetric pseudocapacitors are emerging as a viable commercial alternative for many applications, particularly in the electric mobility market and in grid-scale applications.

5.2.1 Electrochemical Double-layer Capacitors (EDLC)

The mechanism governing the charge storage process of EDLCs relies on the electrochemical double-layer process, *i.e.*, fast adsorption/desorption of charged species at the surface of, typically, high surface area carbon-based materials' electrodes. Capacitance is linked to the potential-dependent accumulation of electrostatic charge at the electrode interface through surface dissociation and ion adsorption, which are the sole means of accumulating surface charge. During the charging process, electrons travel from the negative electrode to the positive one through an external load. Concurrently, cations in the electrolyte migrate towards the negative electrode, and anions move towards the positive electrode, without any change in the electrolyte's composition or concentration.

Owing to the physical nature of the charge storage process of EDLCs, these devices are characterized by a very short time of response (charge–discharge times range from milliseconds to few tens of seconds), which make them the preferred solution to deal with commercial applications where a high-power response is needed. For example, EDLCs can be used in peak shaving in smart grid applications or as backup and safety systems. However, these devices lack energy density, and this fact hampers their widespread use in other applications that are emerging with the energetic transition requiring more energy at high power. Thus, over the last two decades, there has been intensive research to develop new generations of SCs with enhanced energy density at high power density.

A wide range of carbon materials has been studied as electrode materials for EDLCs. Among these, 'supercapacitor grade' coconut carbon is the most renowned. However, its high cost and complex synthesis methods result in a significant carbon footprint and high cost. While other carbon materials are available, their synthesis processes are often costly and energy-intensive, limiting their widespread use. Consequently, alternative options have been explored, with carbon foams emerging as a promising material to enhance the voltage range of EDLCs. For instance, the functionalization of the carbon surface with other atoms or compounds that exhibit faradic responses represents an alternative research area to overcome the modest energy density of carbon materials. For example, doping of carbon with heteroatoms, such as oxygen, nitrogen, boron or sulfur, has been considered a promising route to increase the specific capacitance of EDLCs due to additional redox reactions that such atoms may undergo. These strategies are particularly effective when carefully selecting electrode–electrolyte combinations, as it can significantly boost energy density, which is directly proportional to the square of the cell voltage. The selection of ions in the electrolyte is crucial, as it affects electrolyte conductivity, ion adsorption and dielectric properties, all of which influence the cell's equivalent series resistance and specific capacitance.

In general, commercial carbon SCs have a specific energy ranging from 1 to 10 W h kg^{-1} and a specific power in the range of 500 to 10^4 W kg^{-1}.[5] Their range of applications is highly dependent on the device's final capacitance, which is primarily influenced by the surface area of the carbon material and its pore-size distribution. Since charge storage in EDLCs does not involve chemical reactions, the electrode material does not fade easily, and these devices can sustain up to a few million cycles without relevant capacitance decay. The highly reversible nature of the electrostatic charge storage process reduces possible changes in electrode volume, thereby preventing swelling, mechanical degradation or other issues related to parasitic or irreversible reactions that damage the electrode performance over time. Nevertheless, carbon oxidation has been considered a cause of capacitance fade during the cycling of the device. Also, electrolyte degradation has been considered an important cause of device ageing. Despite all the merits of EDLCs, their energy density is modest, and efforts have been made regarding its increase. Within this context, the introduction of redox species as electroactive electrode materials has been considered the most attractive strategy.

5.2.2 Pseudocapacitors

Pseudocapacitors (or redox supercapacitors) have been considered the next generation of SCs, surpassing EDLCs in terms of energy density without largely compromising high-power metrics. Pseudocapacitors utilize fast electrochemical surface reactions or pseudo-intercalation processes along with electrostatic mechanisms to store charge. This combination allows for increased energy density while maintaining rapid charge–discharge rates. Hence, pseudocapacitors constitute an important solution to fill the gap that exists between EDLCs (high power, low energy) and modern batteries (medium power, high energy). Furthermore, by combining the appropriate materials in the electrode, it is possible to tailor both the power and energy response, establishing an important avenue to design 'fit for the purpose' devices.

Conway[6] identified three main faradaic processes enabling pseudocapacitance (see Figure 5.2): (a) a two-dimensional electrosorption process (so-called underpotential deposition); (b) redox pseudocapacitance; and (c) three-dimensional sorption process with charge transfer (intercalation pseudocapacitance). Despite their similar electrochemical characteristics, these processes rely on distinct physical mechanisms. Underpotential deposition, or reversible adsorption, involves the electrodeposition of a monolayer of metal ions onto another metal surface at potentials above their redox potential. A classic example of underpotential deposition is the deposition of lead on a gold surface[7] (Figure 5.2(a)). Redox pseudocapacitance occurs when ions are adsorbed electrochemically onto a material's surface accompanied by faradaic charge-transfer. The redox pseudocapacitive mechanism is observed, for instance, in RuO_2 film electrodes, with the change of the oxidation state from +4 to +3 (Figure 5.2(b)). On the other hand, intercalation pseudocapacitance involves, as the name suggests, the insertion of ions into tunnels within the electrode material, resulting in a charge-transfer faradaic

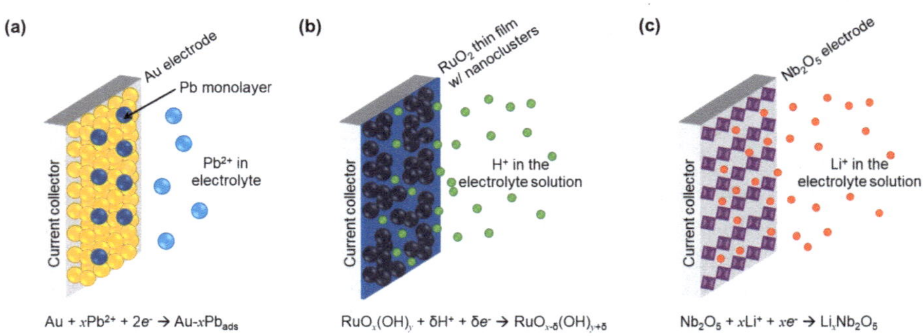

Figure 5.2 Illustration of the three different processes associated with pseudocapacitance: (a) underpotential deposition of lead on the surface of a gold electrode; (b) redox pseudocapacitance of $RuO_2{\cdot}nH_2O$ film electrode; and (c) intercalation pseudocapacitance in a Nb_2O_5 electrode. Reproduced from ref. 7 with permission from the Royal Society of Chemistry.

reaction without a crystallographic phase transition. Currently, niobium pentoxide, T-Nb_2O_5, and MXenes, (*e.g.*, non-oxide transition metal carbides, carbonitrides and nitrides) have been used as pseudo-intercalation materials[8,9] (Figure 5.2(c)).

Since pseudocapacitance relies strongly on faradaic processes, it involves the transfer of charge across the double layer, or even fast intercalation processes, similar to the processes that occur in batteries. In pseudocapacitors, the electrochemical activity takes place at both the electrode's surface/electrolyte interface, and near the bulk of the material, resulting in much higher capacitances compared to EDLCs. Capacitance is determined by the relationship between the charge acceptance of the active material and the change in potential. In pseudocapacitor materials, there is always a contribution from electrostatic double-layer capacitance proportional to their electrochemically accessible interfacial area. The extent of this contribution depends on the material morphology, structure and composition but can vary from a few percent up to 60–70%.

Thus, the charge storage mechanism of pseudocapacitors is always a combination of double-layer and fast redox processes based on reactions occurring at the surface of the electrode. These redox reactions increase the amount of charge that can be stored in the electrode material, making them very attractive for applications that demand increased energy density and fast responses. A pseudocapacitor can reach energy densities that vary from 2 up to 10 times that of a conventional EDLC supercapacitor.

The specific capacitance of pseudocapacitors depends on key factors such as the electroactive material chemical composition, surface area, pore size and pore distribution, structure and overall electrode chemistry. Besides its intrinsic feature towards non-faradaic/faradaic processes, the electrode containing the active material is expected to have high electrical conductivity, large specific surface area, well-tailored porous structure, mechanical, chemical and thermal stability, corrosion resistance, and low cost, and be environmentally friendly.[3]

Reconciling these features is a considerable challenge that creates an extraordinary opportunity to design new materials and architectures, exploring a multitude of chemistries based on transition metal compounds that are electrochemically active in aqueous electrolytes, including low-cost neutral electrolytes. The introduction of metal species in the electrode materials opens a new avenue regarding aqueous based electrolytes. Aqueous electrolytes exhibit high ionic conductivity, reaching nearly 1 S cm^{-1}, which is approximately two orders of magnitude higher than that of organic and ionic liquid electrolytes. Additionally, the hydrated ion sizes in aqueous electrolytes are smaller than those in organic electrolytes, contributing to higher capacity. Aqueous electrolytes are characterized by very low cost, much higher safety, and ease of handling and do not raise concerns when it comes to the recycling processes. However, their working voltage is typically limited within the water splitting window, theoretically 1.23 V. Nevertheless, through functionalization and engineering of the electrode materials, it is possible to increase the overpotential for water splitting and therefore the final voltage window of the device. As a result, aqueous supercapacitors with working voltages as high as 2.0 V have been recently studied using cobalt hydroxide and carbon nanofoams as recently reported.[10]

Traditional electrode architectures are often designed as thin layers on the top of thin metal sheets (the current collector) to minimize the contribution of inactive (or dead) mass, while ensuring the geometrical planar area of the active surface and compactness to increase both gravimetric and volumetric energy density. However, commercial design assemblies do not result in the best electrochemical performance when thin films are used since the charge storage contribution of the active material is compromised by the mass (or volume) of other components that make up the final device.[11] Therefore, a solution to overcome the surface area limitations is to create electrode materials (and also current collectors) with 3D structures that are open and fully accessible to the ions in the electrolyte. This significantly increases the number of active sites participating in the mechanisms responsible for charge storage and, therefore, the capacity of the electrode material. Moreover, these architectures help to lower the total volume or mass of electroactive material needed. In this regard, the use of 3D foams optimizes the use of active materials and has become an interesting solution to create novel electrodes for SCs.

5.3 Carbon Nanofoams

Carbon foams have raised immense commercial interest, including doped carbon foams,[12] for both EDLCs and pseudocapacitors. Carbon foams have been considered of interest, for example, to develop flexible SCs for wearable and portable applications.[13] Surface functionalization of carbon foams with sulfur, oxygen and nitrogen substantially improved the SC electrodes' specific capacitance and achieved very high rate performance and very short time constants.[14] Additionally, carbon foams have been modified or decorated with different transition metal compounds such as manganese oxides (Figure 5.3(a)) or cobalt hydroxides. This option leads to electrodes that may display extended potential windows in aqueous electrolytes and also higher specific capacitances, as evidenced in Figure 5.3(b).

Recently, a carbon nanofoam substrate underwent a functionalization process that, by itself, resulted in a notable 40% increase of the carbon foam's specific capacitance. Subsequently, cobalt hydroxide clusters were formed on the functionalized carbon nanofoam *via* pulse electrodeposition, further enhancing the specific capacitance of the electrode. The synergy of these processes, along with an expanded working potential window, has yielded a material with a specific capacitance nearing 300 F g^{-1} at a current density of 1 A g^{-1}, considering the total mass loading of the composite.[10]

Various methods have been explored for producing carbon nanofoams. One of these methods[15] involves the preparation and characterization of a nanofoam-based supercapacitor electrode material through laser processing

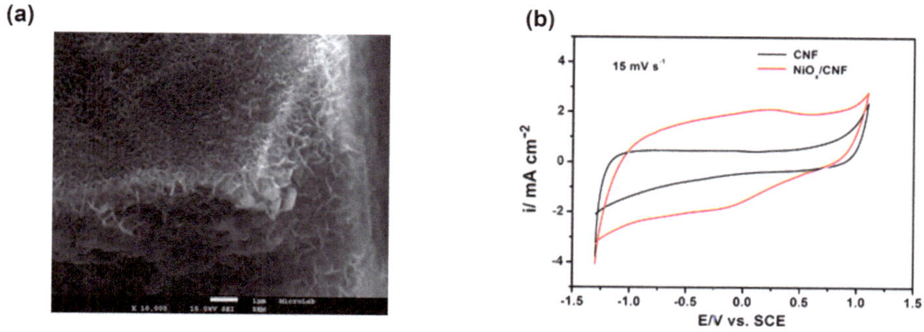

Figure 5.3 (a) Example of carbon foam decorated with thin manganese oxide nanosheets (authors' original picture); and (b) voltammogram obtained on a carbon nanofoam where nickel oxide was electrodeposited, evidencing the extended potential window and enhanced charge storage in Na_2SO_4, an eco-friendly electrolyte.

of a graphene oxide (GO) target. The resulting nanofoams, after being transferred onto a suitable substrate using a water-transfer technique, underwent a morphological transformation that enhanced the specific capacitance of the resulting electrodes by a factor of 2. Another interesting example reported in the literature[16] proposes a supercapacitor cell assembled with a composite electrode made of a carbon nanofoam functionalized with an electrodeposited thin PEDOT film in a nanofibrous structure, alongside microporous PEDOT/graphite asymmetric electrodes approximately 10 μm thick. This device has demonstrated an energy density of 13.6 W h g^{-1} and a high power density of 29.4 kW g^{-1}, which are metrics superior to those obtained for symmetrical supercapacitors based on PEDOT conducting polymers or different structured carbons. The improvement in performance was attributed to the enlarged operating voltages and energy storage mechanisms. Specifically, the carbon nanofoam-fibrous PEDOT composite electrode utilizes an electrical double-layer capacitance mechanism, while the PEDOT/graphite electrodes operate *via* faradic processes.[16]

Using sustainable cotton stalks as a precursor, a carbon nanofoam with remarkable properties has been developed and proposed for a quasi-solid-state supercapacitor.[17] This nanofoam exhibited a large surface area, hierarchical porosity, and a structure doped with oxygen and nitrogen. Notably, it demonstrated high capacitance and high rate capability, even at a high current density of 100 A g^{-1}. This foam material, in an eco-friendly Na/Na_2SO_4 gel electrolyte, displays a wide operating voltage window of 1.8 V, resulting in enhanced energy density of 22.6 W h kg^{-1}. Furthermore, this device performed very well over a 10 000 cycling test.

Hierarchically porous carbon nanofoams with uniform cavities and highly porous skeletons were successfully synthesized *via* a one-pot reaction method.[18] This involved the formation of core–shell organosilica nanoparticles, followed by subsequent Friedel–Crafts chemistry. The level of porosity in these nanofoams was adjusted by varying the quantity of phenyltriethoxysilane used during the initial step of the synthesis process. These nanofoams exhibited remarkable structural features, comprising macroporous cavities with dimensions of approximately 50 nm, mesoporous windows measuring around 12 nm between adjacent cavities, and a mesoporous skeleton with a thickness of approximately 3 nm. Additionally, they possessed exceptional physical properties, evidencing a high surface area exceeding 1100 m^2 g^{-1} and an impressive pore volume of approximately 2 cm^3 g^{-1}. Furthermore, the skeleton of these nanofoams displayed partial graphitization. In terms of their application as supercapacitor electrode materials, these nanofoams demonstrated outstanding performance. Specifically, the nanofoams exhibited a specific capacitance of 170 F g^{-1} at a current density of 0.5 A g^{-1} when tested in an aqueous electrolyte. This excellent electrochemical performance underscores the potential of these hierarchically porous carbon nanofoams for energy storage applications.

Given the attractiveness of graphene as an electroactive material, graphene-based composites displaying a 3D foam architecture have been used as a support for different metals. For example, metallated reduced graphene oxide foam was made from a graphene suspension and a trimetallic (Ni, Co, Zn) zeolitic imidazolate framework (ZIF).[19] A thermal treatment induced the presence of rich metal-doped interfaces. This foam modified with cobalt, nickel and zinc nanoparticles, displayed increased capacitance derived from a very high ionic conductivity in the hierarchical metallated interface. The foam achieved a gravimetric capacitance of approximately 1013 F g^{-1} at 5 mV s^{-1}, retaining nearly 451 F g^{-1} at 100 mV s^{-1}, with 94% retention over 10 000 cycles. The symmetric supercapacitor device attains an energy density of approximately 45 W h kg^{-1} at a power density very close to 750 W kg^{-1}.

The most competitive advantages of these carbon materials include high surface area, tailored porosity and decreased weight. Despite the very attractive features in terms of short time of response and high power, carbon materials show limited capacitance and, as a consequence, modest specific energy. Moreover, it has been reported that the mechanical stability and electrical conductivity of carbon-based materials is somewhat limited for certain applications. To counteract these limitations, metallic foams are commonly selected in industry as current collectors. Metal foams are a creative solution to design supercapacitor electrodes with 3D architectures, not only for their high specific surface area, but also for their great electrical conductivity and simple production methods, as described previously in Chapter 2. The functionalization or decoration of metal foams with electroactive species enhances the electrochemical performance of the electrode towards charge storage and device metrics. Examples of these electrochemically active species are carbon-based materials, that are pressed into the foam porosity, allowing electrolyte penetration and complete wetting of the carbon material; but also different transition metal compounds or carbon–metal compound composites,[20–23] as addressed in Chapter 3.

5.4 Carbon on Metal Foams

As mentioned previously, carbon-based materials can be impregnated onto metal foams, as current collectors. Experimentally, the carbon material is mixed with a binder (usually PVdF) and organic solvents to form an ink-like material that is pressed on the surface of the current collector. Both metal foil and metal foams can be used, not only as current collectors but also as mechanical support of the electroactive carbon material. To enhance the total surface area of the electrode and, consequently, its volumetric energy density, metallic foams are preferred as current collectors. Nickel and aluminium are commonly used due to their good electrical conductivity and relative cost-effectiveness, depending on their compatibility with the electrolyte to be used. There are many examples of devices in which different carbon materials are pressed onto a metallic foam and used as supercapacitor electrodes. Both the carbon material and the porosity/pore size of the metal foam impact on the metrics and electrochemical response of the devices.

In addition to the conventional activated materials, from which the most popular is coconut carbon, other forms of carbon are being researched for supercapacitor electrodes. One of the most widely researched materials in recent years has been graphene, a prime example of a form of carbon vastly used in conjunction with transition metal foams. This combination can be observed in the work of Huang *et al.*[21] with the successful fabrication of functionalized graphene on Ni foam (FG-Ni foam) without the use of any additional conductive agents or binders, highlighting the compatibility of carbon materials with metal foams while lowering the resistance of the FG-Ni foam electrode and improving its cycle life. A similar strategy was described by Yang *et al.*[24] who reported the reduction of graphene oxide by electrochemical routes, in a binder free process, onto a Ni foam surface leading to a porous layer of crumpled nanosheets. The composite material revealed excellent supercapacitive performance that was attributed to the morphological features of the electrodeposited graphene layer and the low contact resistance and increased conductivity in the resulting material.[20,24]

A composite foam made of Ni, Co-MOF-G/Ni foam electrode was fabricated by electrodeposition, revealing a flower-like Ni, Co-MOF within graphene sheets on the Ni foam substrate.[25] These electrodes achieved a specific capacitance of 1158 F g^{-1} at a current density of 2 A g^{-1} in 3 M potassium hydroxide. These interesting values were attributed to the synergistic effects of bi-metal MOFs (Ni, Co-MOF) and porous graphene.

Three-dimensional reduced graphene oxide (RGO) and Ni foam were used to fabricate composite electrodes using a simple and clean method without harmful reducing agents.[20] In this process, the graphene oxide was reduced by the Ni foam in an aqueous suspension at pH 2 at room temperature. The process promoted the growth of reduced graphene oxide sheets around the Ni foam pillars (see Figure 5.4). The areal capacitance could be tuned by varying the reduction times (3 to 15 days). Despite being quite long (several days), the increase of the reduction time leads to a significant increase of

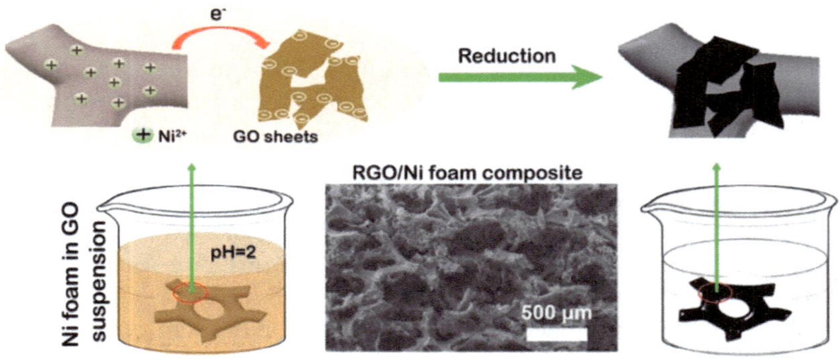

Figure 5.4 Preparation of 3D reduced graphene oxide (RGO)/Ni foam composites. Reproduced from ref. 20 with permission from American Chemical Society, Copyright 2015.

the electrode's areal capacitance from 26.0 to 136.8 mF cm^{-2} at 0.5 mA cm^{-2}. The role of temperature was also studied, and it was shown that it significantly influences reduction efficiency. A composite reduced for 5 hours at 70 °C shows superior properties compared to one reduced for 15 days at room temperature. Extending the reduction to 9 hours at 70 °C results in an areal capacitance of 323 mF cm^{-2} at 0.5 mA cm^{-2}, maintaining good rate performance and cycling stability.[20]

In a very simple and low cost approach, porous graphene networks were electrodeposited onto nickel foams.[24] The electroactive material exhibited a supercapacitive performance resulting from the electrochemical double layer that conferred high rate capability, cycling stability and a specific capacitance close to 183 mF cm^{-2} at 1 mA cm^{-2}.[24] The material revealed a crumpled surface morphology and high conductivity. The graphene sheets were also electrochemically stable on the surface of the 3D foam. This assembly led to short ion diffusion distances, and good contact between the graphene and nickel foam. The study also evidenced that the deposition potential affects the deposition rate, requiring longer times to achieve maximum capacitance under more positive potentials.

In another example, Li *et al.*[26] performed a two-step deposition on top of nickel foam. First, they electrodeposited reduced graphene oxide on the top of a nickel foam. Then, a hydrothermal deposit of MnO_2 was produced to obtain a highly capacitive electrode material with excellent capacitance retention during continuous charge–discharge.[26] Therefore, combining transition metal oxides with carbon-based materials on top of metallic foams is an appealing way of creating the next generation of electrodes for supercapacitors. This strategy explores a synergistic response that involves the classical electrochemical double layer of carbon and the fast and reversible faradic processes that occur in MnOx, a well-known pseudocapacitive electrode material.

More hierarchical combinations can be performed.[27] An interesting example explores the potential of metal chalcogenides and their composites. Rajabathar *et al.*[27] prepared a MoS_2 NPs rolled r-graphene/mesoporous MnOx nanofibre composite using a two-step template-free method. An optimized quantity of MoS_2 particles was added to mesoporous MnOx, and an enhanced thermal stability and electrochemical performance of the electrode was observed. The composite features a nanofibre morphology for MoS_2 rolled graphene sheets. This high-performance supercapacitor electrode material is fabricated on binder-free, flexible nickel foam.

In a similar way, other carbon-based materials have been used as active material for supercapacitors on top of metallic foams. An interesting example was recently reported by Fangya Qi *et al.*[28] who used 'carbon quantum dots' as a kind of glue to prepare honeycomb-like carbon quantum dots in a composite that includes a network of nickel sulfide arrays on a Ni foam. This 3D honeycomb-like architecture is a binder-free electroactive material that exhibits a specific capacitance of 1130 F g^{-1} at 2 A g^{-1} and retains 80% of its initial capacitance at 10 A g^{-1} after 3000 cycles. The material was used to assemble an asymmetrical supercapacitor that displays an energy density of 18.8 W h kg^{-1} at a power density of 134 W kg^{-1} and maintains 100% of its initial capacitance after 5000 cycles at 5 mA cm^{-2}.

A layered material was prepared using nickel foam as the substrate, carbon nanotubes as a structural stabilizer, and Ni–Co bimetallic sulfide as a pseudocapacitive material to create a multilayer composite, NF/CNTs/Ni–Co–S, electrode material.[29] This architecture achieved a specific capacitance of 2952 F g^{-1} at 1 A g^{-1}, retaining 82.9% of its capacitance after 6000 cycles. When assembled into an asymmetric device together with activated carbon – NF/CNTs/Ni–Co–S//AC – it was possible to achieve an energy density of 76.37 W h kg^{-1} at a power density of 799.99 W kg^{-1}.

Other examples of carbon materials that have been combined with metal foams to produce electroactive electrodes include activated carbons obtained from biomass, such as dairy products,[30] cotton[31] or peanuts,[32] carbon fibres[33] or carbon nanotubes,[34] amongst others.

5.5 Transition Metal Oxide Materials (TMOs) on Metal Foams

Nanostructured transition metal foams, particularly Ni foams, have been studied because of their excellent conductivity and stability in different aqueous solutions, and for their employment as substrates of other materials with higher energy storage performances. Both single foams and multi-metal foams have been reported. For example, the substitution of nickel with cobalt, in a foam similar to the one displayed in Figure 5.5, but produced at lower current density, leads to an important change in the fine morphological features that changed from round flower-like to dendrite flower-like, as depicted in Figure 5.5. These foams can be impregnated or electrodeposited with both

Figure 5.5 Nickel foam produced by DHBT, evidencing a porous structure and a flower-like morphology at higher magnifications.

metal oxides, hydroxides or metal sulfides and graphene-based materials, creating electrodes that display a supercapacitive response with high energy density, excellent rate capability and also very good mechanical stability.[35]

Ultrathin Ni–Co double hydroxide (DH) nanosheets were deposited on Ni foam to serve as supercapacitor electrodes as recently reported by Xing *et al.*[36] These Ni–Co DH electrodes, with a Ni/Co ratio of 4/1 in the electrolyte, achieved a specific capacitance of 3028 F g^{-1} at 2 A g^{-1}, along with excellent cyclic stability. This performance was likely attributed to their robust, well-defined nanoporous structure and amorphous nature. Additionally, the Ni–Co DH nanosheets demonstrated an impressive energy density of 127.22 W h kg^{-1} at 0.605 kW kg^{-1}, and they maintained at 93.48 W h kg^{-1} even at a high power density of 15.125 kW kg^{-1}, evidencing their high rate capability.

Cobalt oxide(II, III), Co_3O_4, has been considered an interesting pseudocapacitive material and therefore it has been studied as electroactive electrode material. Recently, three-dimensional (3D) fennel-like Co_3O_4 particles displaying thirty-six surfaces and high surface area were synthesized on nickel foam *via* a hydrothermal route.[37] The unique structure contributed to the excellent electrochemical performance of the electrode. The Co_3O_4 particles on nickel foam demonstrated a high specific capacitance of 384.375 F g^{-1} at a current density of 3 A g^{-1} and exhibited cycling stability with a capacitance retention of 96.54% after 1500 galvanostatic charge–discharge cycles in a 6 M KOH electrolyte. Yu *et al.*[38] also studied Co_3O_4 on Ni foams starting from different precursors that were tested as free-standing electrode materials for supercapacitors. The Co_3O_4 with a thin-nanowire-cluster structure showed the highest electrochemical performance. It delivered a specific area capacitance of 1.92 F cm^{-2} at a current density of 5 mA cm^{-2}, which increased to 2.88 F cm^{-2} after 3000 charge–discharge cycles. Furthermore, it retained a rate capability of 72.91% at a current density of 30 mA cm^{-2}.

Different strategies have been adopted to further increase the response of Co_3O_4 as pseudocapacitive material. Min Kang *et al.*[39] used poly(vinyl pyrrolidone) (PVP) as a directing agent to grow ultrathin Co_3O_4 nanosheets with a highly hierarchical porous structure on a Ni foam. The results demonstrated that such a structure enhances electrolyte diffusion and the electrochemical performance. The electrode achieved a specific capacity of 1196.5 mC cm^{-2} at 5 mA cm^{-2} (798 C g^{-1} at ~3.3 A g^{-1}) and retained around 107% specific capacity after 5000 charge/discharge cycles. An asymmetric supercapacitor made from this electroactive material and activated carbon as negative electrode showed an energy density of 51.7 W h kg^{-1} at 1125 W kg^{-1} and retained 91.8% capacitance after 15 000 cycles.[39]

Another pseudocapacitive material of excellence is RuO_2 that, despite its high price, has been tested as an electrode for supercapacitors. A RuO_2 and graphene composite offering a 'sandwich-like structure' was produced by electrochemical routes under optimized parameters by Li *et al.*[40] This composite offers an accessible surface area where the redox reactions can occur and also stable channels for electrolyte penetration. These features contributed to an enhanced electrochemical performance of the composite. Symmetrical cells assembled with this material exhibited a wide operational voltage window of 1.5 V. The optimized composite demonstrated superior specific capacitance with 89% retention upon increasing the current density by 50 times due to the architecture and contribution of the metal oxide.

The anodic deposition of MnO_x on Ni foams was recently reported by Siwek *et al.*,[41] in which the prepared composites were revealed to be very stable under cycling in alkaline electrolytes. In continuing work, Siwek *et al.*[42] compared the performance of three different Ni-based foams electrodeposited with MnO_2 in an asymmetric device. The authors showed that Ni foam, produced by the DHBT, achieved a larger potential window and, therefore, enhanced energy density compared to Ni–Co foam and commercial Ni foam.

In a recent study, Vainoris *et al.*[43] studied the potentiostatic electrodeposition of Cu to create both 2D plates and Cu foam substrates. The Cu porous structure of the foams contributed with an electrochemically active surface 7 to 14 times higher than the 2D plate, as well as a charge-transfer resistance 1.5–1.7 times lower, practically duplicating the charge-transfer rate by using the 3D structure.[43]

Evidently, TMOs are a group of interesting active materials of varied electrochemical properties that can provide a complementary factor to the response of metal foams *via* simple deposition processes.

Furthermore, both carbon-based materials and metal oxides can be combined to create synergistic materials with enhanced performance. For example, a Ni foam coating with graphene materials obtained by ultrasonic impregnation, followed by electrodeposited porous MnO_2, showcases a creative union of carbon and metal oxides on top of a metal foam.[35]

This strategy enabled a more conductive active electrode material with enhanced charge transfer properties. When assembled in an asymmetric cell, despite the limited potential window (<1 V), the energy density was significantly increased.[35] Another example studied in the literature evidenced the deposition of $ZnCo_2O_4$ nanosheets and graphene, which were deposited by hydrothermal routes onto Ni foam and then MnO_2 was electrodeposited onto the previous composite. When assembled into an asymmetric device it was possible to achieve a 1.6 V window in Na_2SO_4 electrolytes.[44]

Bai *et al.*[45] demonstrated that metal foams can be intermediates to produce carbon foams *via* the process of chemical vapour deposition with methanol, in which Ni foam was used as a template to create graphene foam onto which graphene-coated MnO nanorods were deposited, resulting in a cauliflower-like nanostructure within the already porous graphene 3D framework. The work further associated the capacitance and cycling stability of the resulting electrode to the carbon and metal oxide synergy.[45]

5.6 Summary

The advantages of using metallic foams for the fabrication of SCs bring the possibility to enlarge the active surface area and to create pore distributions that enable faster electrochemical processes. Moreover, these foams can be used as current collectors that can be decorated with other active materials without the need to use additional binders by properly selecting the deposition technique, with electrodeposition or CVD as examples of that feature. The easy functionalization of metallic nanofoams both with metal compounds and transition metal oxides has been considered a step forward in the rational design of SC electrodes tailored for application requirements. Moreover, the active materials used in the composition of the metal foams and/or the coatings result in hybrid behaviours that can display more battery or capacitor-like electrochemical responses, making SCs a broad field of energy storage, flexible enough to address a wide range of applications that address challenges more focused in energy density, power density or a balanced compromise of both.

Nonetheless, there are certain factors that limit the commercial widespread applicability of metallic foams as current collectors for supercapacitors. First, the notable decrease in maximum applicable current remains a challenge. Maintaining the power capabilities is of the upmost importance in supercapacitor applications and most metallic foils, with less porosity, perform at exceedingly higher currents than foams. Moreover, the cost-effectiveness and easy preparation of metallic foils when compared to foams still remains a challenge for the commercialization of foams to be adopted.

It is important to note that there is an immense pool of functionalization strategies of metal foams reported in the literature for supercapacitor applications. However, the electrochemical response of these materials reported as 'supercapacitors' is often closer to a battery-like behaviour instead. This leads to the existence of plateau-shaped curves in galvanostatic charge–discharge curves in most TMO-coated/decorated metal foams (Figure 5.6(a))

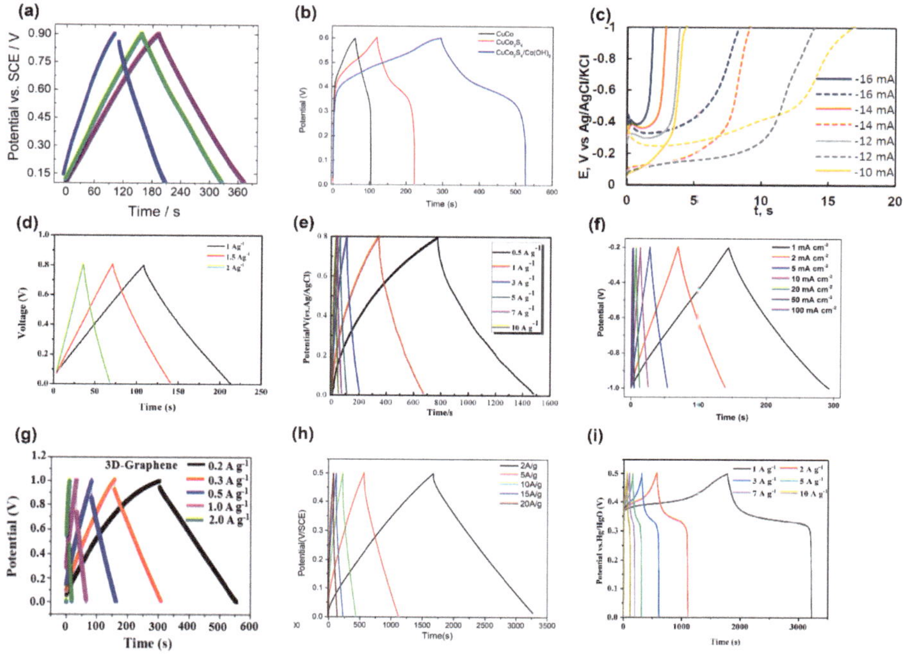

Figure 5.6 Charge–discharge curves of TMO-functionalized metal foams: (a) MnO*x*-based electrodes at 1 A g^{-1} in 1 M Na_2SO_4. Reproduced from ref. 42 with permission from IOP Publishing, Copyright 2022. (b) CuCo, $CuCo_2S_4$ and $Co(OH)_2/CuCo_2S_4$ electrodes at 10 A g^{-1} in 2 M KOH. Reproduced from ref. 46 with permission from Springer Nature, Copyright 2020. (c) On flat (continuous lines) and porous (dashed lines) electrodes at various current densities in 50 mM $CuSO_4$ and 0.45 M Na_2SO_4 solution. Reproduced from ref. 43, https://doi.org/10.3390/coatings10090822, under the terms of the CC BY 4.0 License, https://creativecommons.org/licenses/by/4.0/. Carbon-functionalized metal foams: (d) FG-Ni foam at different current densities in 5 M NaOH. Reproduced from ref. 21 with permission from Elsevier, Copyright 2013. (e) Ni-foam@GNS@MnO_2 electrode in 1 M Na_2SO_4. Reproduced from ref. 26 with permission from Elsevier, Copyright 2015. (f) PG/NF deposited under the potential of −1.2 V with the deposition time of 500 s measured with different current densities in 6 M KOH. Reproduced from ref. 24, https://doi.org/10.1186/1556-276X-9-672, under the terms of the CC BY 4.0 License, https://creativecommons.org/licenses/by/4.0/. Mixed TMO/carbon-functionalised metal foams: (g) 3D-graphene foam measured at different current densities in 1 M Na_2SO_4. Reproduced from ref. 45 with permission from AIP Publishing, Copyright 2020. (h) $MnO_2/ZnCo_2O_4$/G/NF electrodes recorded at different current densities in 1 M Na_2SO_4. Reproduced from ref. 44 with permission from the Royal Society of Chemistry. (i) NMSe/rGO-140 at various current densities in 2 M KOH. Reproduced from ref. 47 with permission from Elsevier, Copyright 2021.

while the carbon and pseudocapacitive TMOs evidence linear-shape curves (Figure 5.6(b) and (c)). Therefore, while TMOs can contribute with a pseudocapacitive response similar to carbon-based materials, most TMOs cannot be considered as pseudocapacitive materials. For them to adhere to the capacitive-like response, the resulting performance of these materials, including the aforementioned composites, must manifest linear-shaped charge–discharge curves as a result of their surface and/or near-surface charge storage mechanism of the electrodes. It is also important to note that the metallic foams used as current collectors do not present faradaic reactions that interfere with the electrochemical signal from the active material, while ensuring that they have a long lifetime, that is, they do not present any degradation in the electrolyte that could compromise the life of the supercapacitor. Figure 5.6 shows some examples of charge–discharge curves of TMO-functionalized metal foams,[42,43,46] carbon-functionalized metal foams[21,24,26] and mixed TMO/carbon-functionalized metal foams.[44,45,47]

References

1. Q. Li, B. Lin, S. Zhang and C. Deng, *J. Mater. Chem. A*, 2016, **4**, 5719–5729.
2. R. Li, S. Wang, Z. Huang, F. Lu and T. He, *J. Power Sources*, 2016, **312**, 156–164.
3. V. H. Nguyen and J.-J. Shim, *J. Power Sources*, 2015, **273**, 110–117.
4. C. Yuan, J. Li, L. Hou, X. Zhang, L. Shen and X. W. D. Lou, *Adv. Funct. Mater.*, 2012, **22**, 4592–4597.
5. P. Simon and Y. Gogotsi, *Nat. Mater.*, 2020, **19**, 1151–1163.
6. B. E. Conway, *Electrochemical Supercapacitors Scientific Fundamentals and Technological Applications*, Springer, New York, NY, 1999.
7. V. Augustyn, P. Simon and B. Dunn, *Energy Environ. Sci.*, 2014, **7**, 1597–1614.
8. K. Li, X. H. Wang, X. F. Wang, M. Y. Liang, V. Nicolosi, Y. X. Xu and Y. Gogotsi, *Nano Energy*, 2020, **75**, 8.
9. F. Shen, Z. T. Sun, Q. G. He, J. Y. Sun, R. B. Kaner and Y. L. Shao, *Mater. Horiz.*, 2021, **8**, 1130–1152.
10. R. Della Noce, S. Eugénio, T. M. Silva, M. J. Carmezim and M. F. Montemor, *J. Power Sources*, 2015, **288**, 234–242.
11. H. Sun, J. Zhu, D. Baumann, L. Peng, Y. Xu, I. Shakir, Y. Huang and X. Duan, *Nat. Rev. Mater.*, 2018, **4**, 45–60.
12. J. Wang, L. Shen, P. Nie, X. Yun, Y. Xu, H. Dou and X. Zhang, *J. Mater. Chem. A*, 2015, **3**, 2853–2860.
13. M. Cao, Y. Feng, R. Tian, Q. Chen, J. Chen, M. Jia and J. Yao, *Carbon*, 2020, **161**, 224–230.
14. H. Peng, B. Yao, X. Wei, T. Liu, T. Kou, P. Xiao, Y. Zhang and Y. Li, *Adv. Energy Mater.*, 2019, **9**, 1803665.
15. S. Nufer, P. Lynch, M. Cann, M. J. Large, J. P. Salvage, S. Víctor-Román, J. Hernández-Ferrer, A. M. Benito, W. K. Maser, A. Brunton and A. B. Dalton, *ACS Omega*, 2018, **3**, 15134–15139.

16. N. K. Sidhu and A. C. Rastogi, *Mater. Chem. Phys.*, 2016, **176**, 75–86.
17. Z. W. Li, S. Gao, H. Y. Mi, C. C. Lei, C. C. Ji, Z. L. Xie, C. Yu and J. S. Qiu, *Carbon*, 2019, **149**, 273–280.
18. Y. Zhang, Y. F. Wang and A. G. Hu, *RSC Adv.*, 2015, **5**, 70297–70301.
19. R. Patil, L. Pradhan, B. M. Matsagar, R. R. Salunkhe, K. C.-W. Wu, B. K. Jena and S. Dutta, *RSC Appl. Interfaces*, 2024, **1**, 483–491.
20. J. Yang, E. Zhang, X. Li, Y. Yu, J. Qu and Z. Z. Yu, *ACS Appl. Mater. Interfaces*, 2016, **8**, 2297–2305.
21. Z. Huang, H. Zhang, Y. Chen, W. Wang, Y. Chen and Y. Zhong, *Electrochim. Acta*, 2013, **108**, 421–428.
22. G. Ren, X. Pan, S. Bayne and Z. Fan, *Carbon*, 2014, **71**, 94–101.
23. S. Ye, J. Feng and P. Wu, *ACS Appl. Mater. Interfaces*, 2013, **5**, 7122–7129.
24. S. Yang, B. Deng, R. Ge, L. Zhang, H. Wang, Z. Zhang, W. Zhu and G. Wang, *Nanoscale Res. Lett.*, 2014, **9**, 672.
25. S. Salehi and I. Karimzadeh, *J. Prog. Phys. Appl. Mater.*, 2023, **3**, 211–216.
26. Y. Li, D. Cao, Y. Wang, S. Yang, D. Zhang, K. Ye, K. Cheng, J. Yin, G. Wang and Y. Xu, *J. Power Sources*, 2015, **279**, 138–145.
27. J. R. Rajabathar, H. A. Al-lohedan, P. Arunachalam, Z. A. Issa, M. K. Gnanamani, J. N. Appaturi, S. N. Ibrahim and W. Mohammed Dahan, *J. Alloys Compd.*, 2021, 850.
28. F. Y. Qi, L. Y. Shao, X. Y. Shi, F. D. Wu, H. T. Huang, Z. P. Sun and A. Trukhanov, *J. Colloid Interface Sci.*, 2021, **601**, 669–677.
29. H. Liu, S. L. An, X. J. Sun, X. X. Han, J. L. Cui, Y. Q. Zhang and W. X. He, *Colloids Surf., A*, 2021, **629**, 8.
30. K. I. Popov, S. S. Djokić, N. D. Nikolić and V. D. Jovic, *Morphology of Electrochemically and Chemically Deposited Metals*, Springer, Switzerland, 2016.
31. M. Vijayakumar, A. B. Sankar, D. S. Rohita, T. N. Rao and M. Karthik, *ACS Sustainable Chem. Eng.*, 2019, **7**, 17175–17185.
32. N. T. Nguyen, P. A. Le and V. B. T. Phung, *Chem. Eng. Sci.*, 2021, **229**, 8.
33. D. Sridhar, J. L. Meunier and S. Omanovic, *Mater. Chem. Phys.*, 2019, **223**, 434–440.
34. D. D. Zhao, Z. Yang, L. Y. Zhang, X. L. Feng and Y. F. Zhang, *Electrochem. Solid-State Lett.*, 2011, **14**, A93–A96.
35. R. Liu, R. Jiang, Y. H. Chu and W. D. Yang, *Nanomaterials*, 2021, **11**, 2736.
36. J. H. Xing, S. Y. Wu and K. Y. S. Ng, *RSC Adv.*, 2015, **5**, 88780–88786.
37. Y. F. Li, Z. Y. Hai, X. J. Hou, H. Y. Xu, Z. X. Zhang, D. F. Cui, C. Y. Xue and B. Z. Zhang, *J. Nanomater.*, 2017, **2017**, 8.
38. Z. Y. Yu, Z. X. Cheng, Z. X. Tai, X. L. Wang, C. M. Subramaniyam, C. S. Fang, S. Al-Rubaye, X. T. Wang and S. X. Dou, *RSC Adv.*, 2016, **6**, 45783–45790.
39. M. Kang, H. Zhou, P. S. Wen and N. Zhao, *ACS Appl. Energy Mater.*, 2021, **4**, 1619–1627.
40. M. Li and H. W. He, *Appl. Surf. Sci.*, 2018, **439**, 612–622.
41. K. I. Siwek, S. Eugénio, T. M. Silva and M. F. Montemor, *Electrochim. Acta*, 2018, **281**, 39–47.

42. K. I. Siwek, S. Eugénio, I. Aldama, J. M. Rojo, J. M. Amarilla, A. P. C. Ribeiro, T. M. Silva and M. F. Montemor, *J. Electrochem. Soc.*, 2022, **169**, 020511.
43. M. Vainoris, H. Cesiulis and N. Tsyntsaru, *Coatings*, 2020, **10**, 822.
44. W. J. Cong, R. Miao, B. R. Tao and F. J. Miao, *RSC Adv.*, 2019, **9**, 32889–32897.
45. X.-L. Bai, Y.-L. Gao, Z.-Y. Gao, J.-Y. Ma, X.-L. Tong, H.-B. Sun and J. A. Wang, *Appl. Phys. Lett.*, 2020, **117**, 183901.
46. F. Miao, X. Li, B. Tao and P. K. Chu, *Ionics*, 2020, **26**, 5241–5249.
47. G. Zhang, H. Xuan, J. Yang, R. Wang, Z. Xie, X. Liang, P. Han and Y. Wu, *J. Power Sources*, 2021, **506**, 230255.

CHAPTER 6

Metallic Foams for Modern Battery Applications

6.1 Introduction

Modern batteries indisputably play a key role in the current transition towards more sustainable energy and are essential to power portable systems and enable electric mobility.[1] Among other features, batteries exhibit high energy density in mid- to long-term use, low self-discharge, long-term stability and suitable reliability, which highlights their impact on the next generation of electrochemical energy storage devices. Batteries store energy by chemical reactions based on charge transfer reactions with ion (de)intercalation processes with changes in the oxidation state and phase transitions in the electrode material. Once these electrodes are connected externally, these reactions occur in tandem at both electrodes, and electrons are released, providing current to be used. Surface and bulk redox reactions enable storing high amounts of energy, although their power response is modest and their life-cycle shorter compared to other technologies, such as supercapacitors. Owing to these advantages, interest in novel materials for battery electrodes has been growing over the past few years and is currently a hotspot of academic research worldwide. In the following sections, the application of transition metal foams in different battery systems is discussed.

Transition Metal-based Nanofoams for Electrochemical Systems: Manufacturing, Characterization and Applications
By Maria de Fátima Montemor, Diana M. Fernandes, Alberto Adan-Mas, Ana Catarina Alves, Gabriel Garcia Carvalho and Inês S. Marques

Published by the Royal Society of Chemistry, www.rsc.org

6.2 Lead–acid Batteries

Amongst battery technologies, lead–acid (Pb–acid) batteries have dominated the market since the late 19th century. Nowadays, Pb–acid batteries still account for above 70% of the total battery market, despite the recent increase in demand of Li-ion batteries, including transportation, emergency systems, stationary storage and communications.[2] The popularity of Pb–acid batteries mainly relies on their low cost and availability, as well as the easy recycling of lead metal.[3] In fact, nowadays, recycled lead and recycled plastic are used in Pb–acid batteries for electrodes and casing, respectively, and many production facilities integrate recycling units. Pb–acid batteries are strongly implemented in the market and are of outmost importance for the energy transition, and a vital component in the automotive industry and stationary applications. However, Pb–acid batteries have relatively modest energy density (~30 to 40 kW h kg^{-1}, see Figure 6.1), limited power and cycling, and lead corrosion issues, which limits their range of applications. Foams have been used as an alternative to overcome the low energy density. Therefore, porous carbon materials or metal foams have been used as current collectors, coated with an electrochemically active transition metal to produce the electrode. This strategy of using carbon-based foams as current collectors for Pb–acid batteries has been proposed and documented in several patent applications.[4–7] The importance of the use of a high surface area carbon material as current collector is related to several parameters: (1) heavy weight of lead grids that constrain the energy density of the battery, (2) the significant volume expansion (4 to 7%[5]) of the electrodes during discharge processes, and (3) corrosion resistance of carbon. All of these issues will have an impact on the energy efficiency of the energy storage system.

In Pb–acid batteries, carbon-based foams are directly used as negative current collectors, and lead, lead-alloy foams and copper foams structures have been used as positive current collectors.[2,8–16] Despite their high recyclability

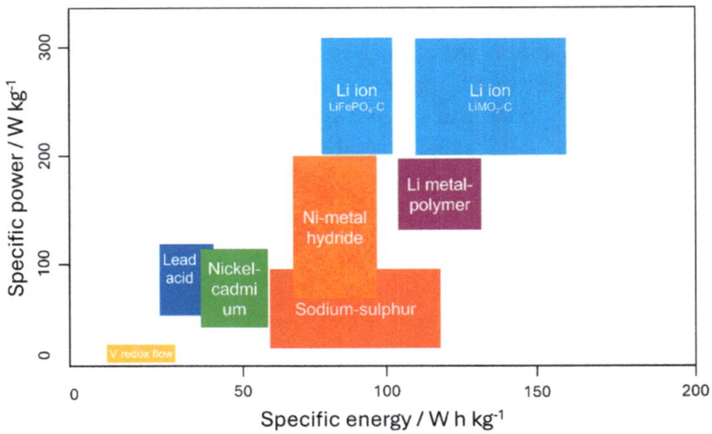

Figure 6.1 Gravimetric power and energy densities for different rechargeable batteries.[18]

and widespread adoption, Pb–acid batteries are somewhat limited in their power capabilities. Moreover, lead is a poisonous heavy metal that is environmentally hazardous, affecting millions of people per year.[17,18] This has prompted battery researchers to strive for greener alternatives without compromising the energy storage performance and affordability of the device.

6.3 Lithium-ion Batteries

Currently, lithium-ion (Li-ion) batteries play a major role in the transition towards increased electrification, being one of the fastest growing markets. Li-ion batteries' fast growth is associated with their significantly higher energy and power density compared to Pb–acid batteries. However, there is a pressing need for material developers to find solutions to the increasing price of the raw metals used in these batteries. In modern Li-ion batteries, the cathode is composed of a Li-containing metal oxide ($LiMO_2$, M = Co, Li, Mn[18,19]) or lithium iron phosphate ($LiFePO_4$),[18] while the anode material varies from the conventional graphite to more advanced Si-graphite composites (Figure 6.2).[20] The mechanism of charge storage in Li-ion batteries follows the concept of a 'rocking-chair': a reversible intercalation/deintercalation of lithium ions within the bulk electrode structure, shuttling forward and back between the two electrodes. The following general reactions translate the charge (eqn (6.1)) and discharge (eqn (6.2)) mechanisms when a lithium metal oxide cathode and a conventional graphite anode are used:

$$\mathrm{LiMO_2} \rightleftharpoons \mathrm{Li_{(1-x)}MO_2} + x\mathrm{Li^+} + xe^- \quad \textbf{Cathode} \tag{6.1}$$

$$6\mathrm{C} + x\mathrm{Li^+} + xe^- \rightleftharpoons \mathrm{Li_xC_6} \quad \textbf{Anode} \tag{6.2}$$

$$\mathrm{LiMO_2} + 6\mathrm{C} \rightleftharpoons \mathrm{Li_{(1-x)}MO_2} + \mathrm{Li_xC_6} \quad \textbf{Overall reaction} \tag{6.3}$$

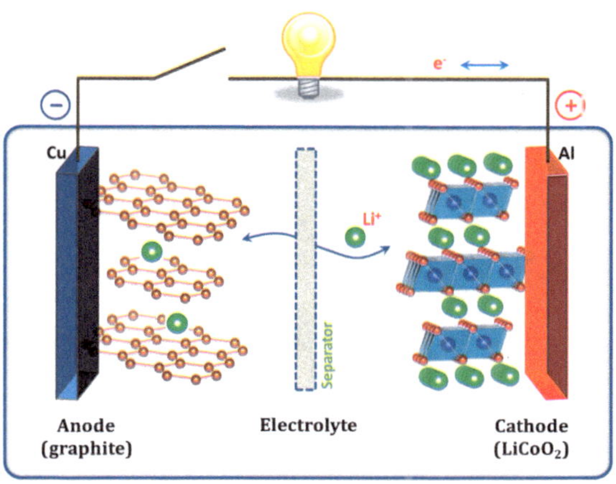

Figure 6.2 Schematic illustration of a Li-ion battery. Reproduced from ref. 20 with permission from American Chemical Society, Copyright 2013.

As a result of the increased demand for Li-ion batteries, researchers are facing the challenge of rationally designing electrode materials with enhanced metrics, efficiency and effectiveness, without compromising the electrochemical and mechanical features. Besides that, the development of new anode material for Li-ion batteries is mandatory to replace metallic Li anodes that suffer from extreme volume expansion and dendrite grown and, hence, affect the security and function of the device that limit the applicability of the battery.

Carbon-based anodes have been widely used as electrodes in Li-ion batteries. Owing to the intercalation/deintercalation of Li^+ ions between different planes on the carbon structure, the electrochemical performance of carbon-based materials is ensured by the mechanical stability of the structure and high electrical and ionic conductivity.[21,22] Besides being extremely abundant in nature, carbon-based materials present low lithiation potential and minimal volume changes during the charge/discharge processes. Lithium-intercalated graphite is essential in LIBs and therefore different types and sources of carbon-based materials have been extensively studied for application in Li-batteries. However, commercial graphite anodes exhibit modest theoretical capacity (372 mA h g^{-1} [22,23]), which limits the battery capacity and the high energy density requirements, as shown in Figure 6.3. In fact, currently, other materials such as silicon or different metal oxides and alloys have been regarded as alternatives for LIB anodes. These materials typically exhibit relatively low delithiation potential and increase the anode capacity making them very attractive electrodes for Li-ion batteries.

Current collectors have a significant role in the structure of an energy storage system. In commercial Li-ion batteries, the coating containing the active material is placed into quasi-two-dimensional (2D) metal foil current collector substrates that are extremely thin, typically 50 to 100 µm, to sustain high specific (or volumetric) capacity. By increasing the total energy available per cell, the battery pack will require a lower number of cells and, therefore, the

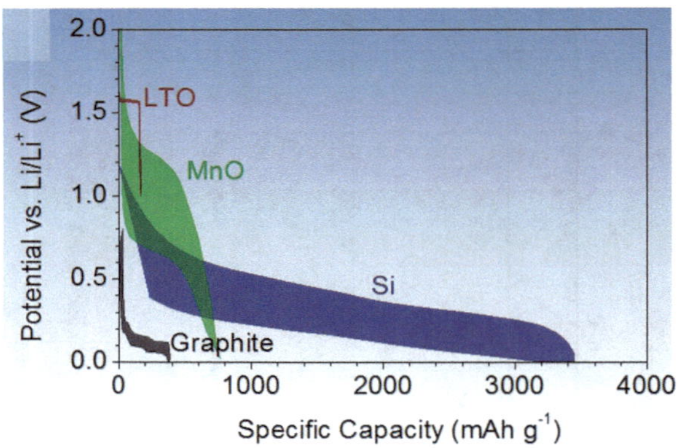

Figure 6.3 Charge and discharge profiles of graphite, lithium titanium oxide (LTO), manganese(II) oxide (MnO) and silicon. Reproduced from ref. 21 with permission from Elsevier, Copyright 2015.

overall weight and volume of the battery pack will be decreased. However, physical damage, such as formation of cracks and defects in the electrodes' coatings, may compromise the reliability of the battery and, therefore, alternative solution materials have been considered, rather than conventional dense metal foils.[24,25] In this sense, a possible solution is to homogeneously apply the active materials onto a three-dimensional (3D) porous metal current collector surface, enhancing the mass loading and electron conductivity, and preventing mechanical issues.[24,26] Metal foams have been used as current collectors in Li-ion batteries so the micro/nanoporous structure of foams can accommodate more active material using thinner layers, allowing the creation of more reaction sites and reducing the diffusion path of ions for redox reactions. Several metal foams have been applied to Li-ion batteries, particularly Ni-[25,27–35] and Cu-based[23,36–47] foams.

6.3.1 Ni-based Foams in Li-ion batteries

Commercial Ni-based foams are nowadays available for Li-ion battery technologies, because of their excellent conductivity and stability in different neutral, acid and alkaline aqueous solutions. For increasing the electrochemical performance of batteries, Ni foams are also excellent substrates for the deposition of other materials (transition metals or carbon materials) with higher energy storage performance. These novel advances on Ni-based foam electrodes are essential for enhancing, for instance, the micromechanical properties and compressive behaviour of foams in automotive applications.[32] The 3D high surface area and open porosity enhances electrolyte transport, the ionic transport process and reduces the Li-ion diffusion path.

Ni-based materials have been used as either anode[25,27–30,48–57] or cathode[31,33,58,59] electrodes of Li-ion batteries. Particularly, NiO-based materials have been investigated as anodes because of their environmental abundancy, non-toxicity and increased volumetric energy density.[34] Compared to other nanoporous materials, Ni oxides attain a theoretical capacity of 700 mA h g^{-1},[34] a much higher value than that of commercial carbon-based anodes. The charge storage mechanism in NiO involves the lithiation and delithiation (intercalation/deintercalation, respectively) of the Li^+ ion, with the formation and deformation of lithium oxide (Li_2O) as the nickel composite is reduced, according to the following chemical reaction[34,60] proposed by Tarascon *et al.* for a transition metal oxide:

$$MO + 2Li^+ + 2e^- \rightarrow Li_2O + M \quad (6.4)$$

$$2Li \leftrightarrow 2Li^+ + 2e^- \quad (6.5)$$

Overall:

$$MO + 2Li \leftrightarrow Li_2O + M \quad (6.6)$$

$$Li_2O + M \rightarrow MO + 2Li^+ + 2e^- \quad (6.7)$$

This reaction can be schematically represented by Figure 6.4.

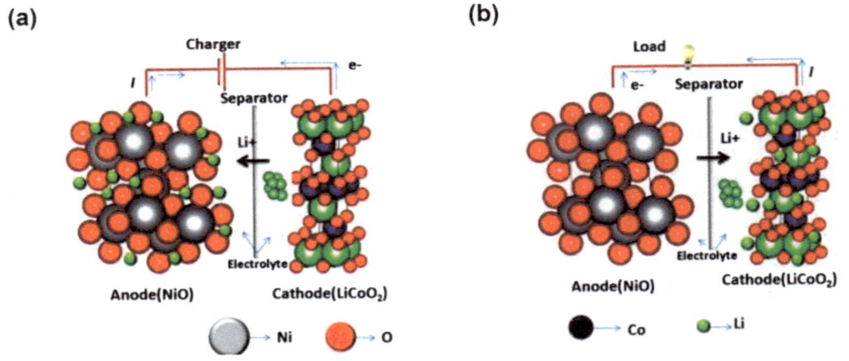

Figure 6.4 Schematic representation of (a) the lithiation (charge), and (b) delithiation (discharge) using NiO as the anode and $LiCoO_2$ as the cathode. Reproduced from ref. 34 with permission from John Wiley & Sons, Copyright 2022.

Despite their high theoretical capacity, NiO materials are characterized by low reversible capacity due to low electronic conductivity.[48] In this way, the preparation of NiO-based electrodes with improved electrochemical performance was investigated using different approaches that enabled various metrics. Huang *et al.*[54] developed Ni-foam-supported NiO and NiO/polyaniline (PANI) films using a chemical bath deposition technique, obtaining nanoflakes that are further coated with PANI. The NiO/PANI film exhibited higher specific capacity (520 mA h g^{-1}@1C) in relation to that of the NiO film (440 mA h g^{-1}), which was attributed to the enhanced electrical conductivity of PANI. Taking advantage of a thermal treatment, Wang *et al.*[49] obtained porous NiO on Ni foam capable of achieving a high specific capacity of 520 mA h g^{-1} at 0.2C with no noticeable capacity fade up to 30 cycles and phase change during consecutive charge/discharge processes. Yang *et al.*[53] also prepared a NiO nanorod anchored on Ni foams using thermal annealing, obtaining 802.4 mA h g^{-1} at 1.4C (1C = 718 mA g^{-1}[48]) and 71.2% coulombic efficiency. However, an irreversible capacity during the first charge/discharge cycle was observed and attributed to the formation of a solid electrolyte interface (SEI) layer and incomplete deformation of Li_2O (see eqn (6.4)), as shown in Figure 6.5(a). In the following cycles, >98% coulombic efficiency was obtained, demonstrating long-term stability. Moreover, when cycled at different current densities, the NiO/Ni foam electrode exhibited reversible capacities and decreased regularly with the increase of the current. When the current density is returned to lower values (0.5 A g^{-1}), the electrode recovered its capacity, indicating good structural stability and rate performance (see Figure 6.5(b)).

The preparation of NiO films on Ni-foams can be achieved by different routes. As reported by Wang *et al.*,[48] a NiO film on Ni foam was prepared by electrodeposition and thermal treatment, in which the anode achieved a capacity of 651 mA h g^{-1} at 0.1C with 93.2% capacity retention.[48] Instead

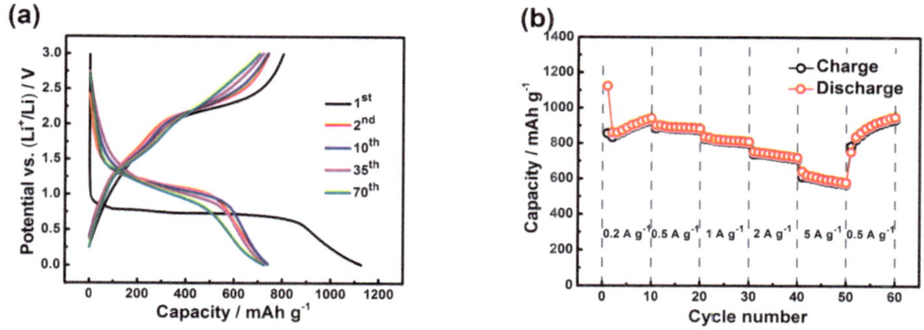

Figure 6.5 (a) Galvanostatic charge/discharge curves of NiO/Ni foam in the potential range of 0.01 to 3 V (*vs.* Li^+/Li), in different stages of charge and discharge, at 1 A g^{-1}. (b) Rate performance of NiO/Ni foam at different current densities. Reproduced from ref. 53 with permission from the Royal Society of Chemistry.

of an electrodeposited material, Ni *et al.*[50] used chemical liquid deposition followed by an annealing treatment to prepare the NiO/Ni foam electrodes. The formation of a SEI layer was also assumed in this work, as the initial discharge capacity is much higher than the subsequent charge.[50] NiO/Ni foams exhibited a 1.38 mA h cm^{-2} capacity at 0.4C that was maintained after 140 cycles, assigned to the porous structure and to the lack of aggregation of particles during cycling. In the work developed by Song *et al.*,[51] NiO/Ni foams were fabricated using freeze/drying casting followed by a thermal oxidation of commercially Ni foams, and applied as anodes to a Li-ion coin cell battery. The NiO/Ni foams demonstrated a high-rate performance (861 mA h g^{-1}) compared to a conventional Ni foam (70 mA g^{-1}), and a stable cycling performance, achieving a charge capacity of 500 mA h g^{-1} after 50 cycles, due to a large surface area and uniform structure. After lithiation of NiO/Ni foams, the crystalline nature of the material was converted into an amorphous form.

As previously mentioned, the most well-known and commercially available cathodes for Li-ion batteries are lithium metal oxides that include, for example, $LiCoO_2$, $LiNiO_2$, $LiMnO_4$ and $LiFePO_4$ materials (Figure 6.6(a)).[18,31] The major advantage of these materials is their intercalating ability for Li ions from the electrolyte. $LiCoO_2$ was introduced by Goodenough in 1980[61] and further commercialized by SONY owing to its high theoretical specific capacity of 274 mA h g^{-1}, low self-discharge and good cycling performance.[21] However, these electrodes are not only expensive, because of the high cost of cobalt, but also present low thermal stability and might have a negative impact on the environment if not properly recycled. Despite that, $LiCoO_2$ electrodes are still used in commercial Li-ion batteries.

Exploring polyanion-containing cathode materials, $LiFePO_4$-based (LFP) electrodes exhibit a high theoretical capacity of 170 mA h g^{-1}, low cost and good cycling performance.[33] However, their major weakness is the low

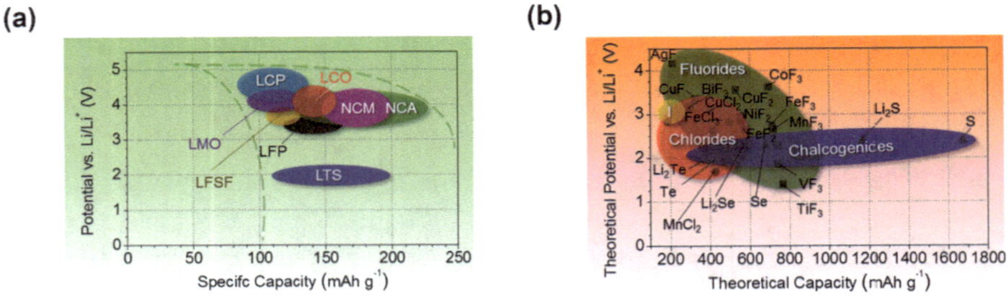

Figure 6.6 Discharge potentials and specific capacity of (a) some of the most common intercalation-type cathodes; and (b) conversion-type cathodes. Reproduced from ref. 21 with permission from Elsevier, Copyright 2015. The acronyms are: LCO = $LiCoO_2$, LMO = $LiMnO_2$, NCM = $LiNi_xCo_yMn_zO_2$, NCA = $LiNi_{0.8}Co_{0.15}Al_{0.05}O_2$, LCP = $LiCoPO_4$, LFP = $LiFePO_4$, LFSF = $LiFeSO_4F$ and LTS = $LiTiS_2$.

average potential window and poor electrical conductivity (~10^{-9} cm S^{-1}), which was the driving force for the development of $LiFePO_4$-based composites with superior performance. Their inherent safety is also a major driver for the commercial adoption of LFP.

The introduction of 3D metal foams can be used to enhance the surface area of the electrode and decrease the diffusion length of Li ions, improving electrical conductivity.[33] For instance, Song *et al.*[33] prepared a Ni-based foam alloy (NiCrAl metal foam) with different contents of carbon (5 to 20 wt%). Electrodes containing 15 and 20 wt% demonstrated a similar electrochemical performance and compared with electrodes with lower percentage of carbon, showed a higher capacity value 150 mA h g^{-1} at 0.2C. Despite that, these electrodes displayed higher electrode polarization due to higher over-potential, therefore the optimal content of carbon was described to be dependent on the application of the electrode.[33]

Other types of cathodes used in Li-ion batteries are known as conversion electrodes that undergo a solid-state reaction during the intercalation/deintercalation of Li ions into the structure. In this case, there is a change to the crystalline structure of the material, in which reversible breaking and recombination of chemical bonds occur during the charge and discharge process. Discharge potentials and specific capacity of examples of these electrodes are depicted in Figure 6.6(b). Metal sulfides, in particular Ni_xS_y materials, are promising electrodes because of their high theoretical capacity (up to 590 mA h g^{-1}), chemical stability and natural abundance.[62] In order to improve cycling performance and conductivity, the preparation of Ni_xS_y materials on Ni foam substrates as electrodes has been studied and reported in the literature.[31,58,59] For instance, Li *et al.*[58] prepared a template-free hydrothermal Ni_2S_3 nanotube array on Ni foam, in which the electrode material exhibited 762 mA h g^{-1} at 100 mA g^{-1} at the 100th cycle of charge/discharge (Figure 6.7(a)). Although the increase of the current

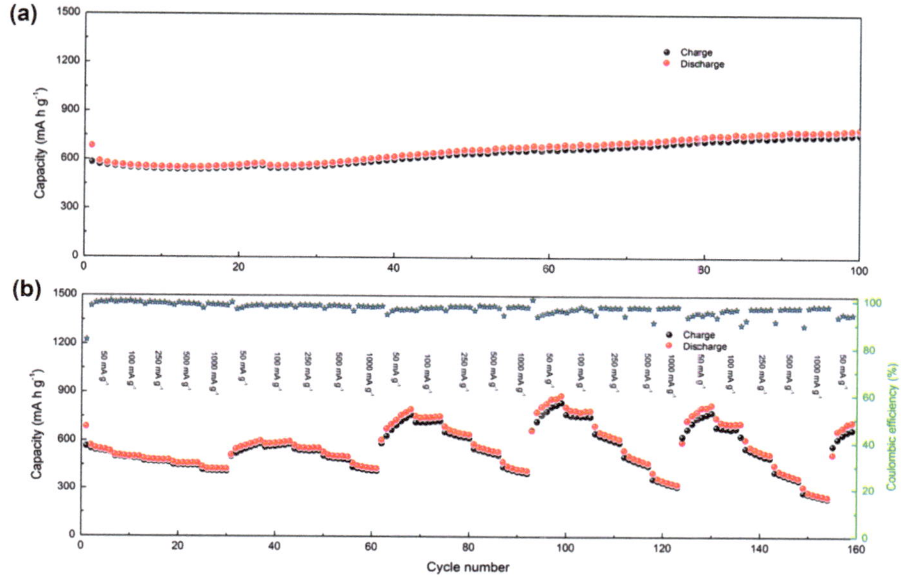

Figure 6.7 Electrochemical behaviour of Ni_3S_2 nanotube array electrodes on Ni foam in the potential range of 0.02–3.0 V (*vs.* Li/Li^+). (a) Charge/discharge cycling of the cathode initially at 50 mA g^{-1} (for 22 cycles), then increased to 100 mA g^{-1} for 80 cycles. (b) Rate performance of the electrode at several scan rates from 50 to 1000 mA g^{-1}. Reproduced from ref. 58 with permission from the Royal Society of Chemistry.

density to 1 A g^{-1} decreases the specific capacity value to 419 mA h g^{-1}, a satisfactory reversible capacity behaviour was observed, achieving 532 mA h g^{-1} after 160 cycles, as shown in Figure 6.7(b). In another work, Feng *et al.*[31] developed a Ni_2S_3 film on Ni foam, also by hydrothermal processes, in which the Ni foam was used as precursor reactant and substrate for the reaction. The electrochemical performance of these cathodes as the electrode for Li-ion batteries was evaluated according to the film thickness (480, 630, 670 and 920 nm), in which the Ni_2S_3-670 nm sample exhibited an initial discharge and charge capacities of 596 and 466 mA h g^{-1} (at 50 mA g^{-1}), respectively. The low coulombic efficiency of these species was then associated with the formation of a SEI layer on the nanostructure. When the current density increased from 0.2 to 5 mA cm^{-2}, the Ni_3S_2 film, with a thickness of 670 nm, showed 64% capacity retention, as shown in Figure 6.8(a). Moreover, when compared to other samples, the Ni_2S_3-670 nm showed better cycle stability, reaching 421 mA h g^{-1} capacity after 60 cycles of charge/discharge, according to Figure 6.8(b). Such a result demonstrated the influence of the film thickness on the kinetics of the charge storage reaction.

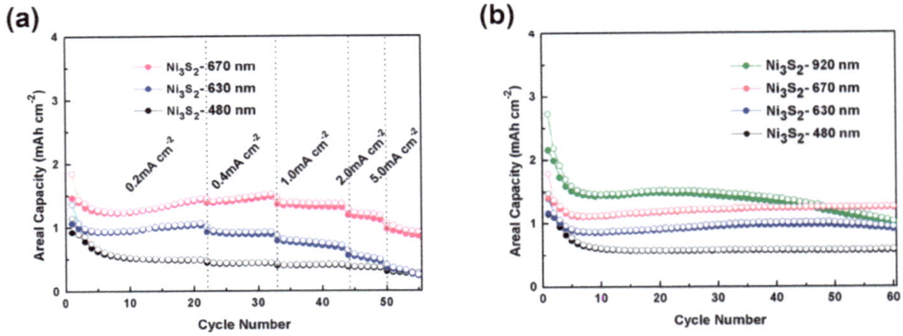

Figure 6.8 (a) Rate capability of the electrodes with different Ni_3S_2 film thicknesses at different scan rates (0.2 to 5 mA cm^{-2}) in a voltage range of 0.01–3.0 V (*vs.* Li/Li$^+$); and (b) capacity retention of the galvanostatic tests for the electrodes with different Ni_3S_2 film thicknesses at 50 mA g^{-1} for 60 consecutive charge/discharge cycles, between 0.01–3.0 V (*vs.* Li/Li$^+$). Reproduced from ref. 31 with permission from the Royal Society of Chemistry.

6.3.2 Cu-based Foams

Copper foams, which are very simple to obtain, have been considered an attractive electrode material for current collectors for Li-ion batteries.[31,33,36–43,58,59] The use of Cu foams seems to mitigate the growth of Li dendrites and contribute to the enhanced rate capability due to their high surface area. In a recent study, Vainoris *et al.*[43] studied the potentiostatic electrodeposition of Cu to create both 2D plates (plane/smooth foil) and Cu foam substrates. The Cu porous structure of the foams contributed to an electrochemically active surface 7 to 14 times higher than a conventional 2D sheet, as well as a charge-transfer resistance 1.5 to 1.7 times lower, practically duplicating the charge-transfer rate as a result of the 3D structure of the collector.[43]

Among other materials, tin (Sn) is considered as a material with high potential for replacing graphite in Li-ion batteries, owing to its high theoretical capacity (991 mA h g^{-1}). However, pure Sn anodes suffer from large volumetric changes during the lithiation/delithiation process, which leads to delamination of the active material from the current collector. Porous Cu foams have also been widely used as substrates for Sn-based anodes for Li-ion batteries,[41,43] and the use of Cu foams in Sn-based anodes was noted as essential for the successful fabrication of the electrode.[37,39,40] Li *et al.*[40] prepared a porous Sn film (of 0.6 μm average thickness) using electroless deposition on Cu foams, obtaining an initial capacity of 676 mA h g^{-1} at 50 mA g^{-1}, and an excellent cycling performance (313 mA h g^{-1} after 100 cycles of charge–discharge) compared to other thicker films. The authors explained that this observation is a result of the smaller volume expansion and stronger adhesion of the thinner electrode to the foam. The capacity drop after the initial charge (Li$^+$ deintercalation) was also observed by Kim *et al.*,[39] due to the reduction of the oxide in the anode (as discussed in the previous section).

Moreover, similar values were obtained in this study, achieving an initial capacity of 567 mA h g^{-1}, that then decreased to 340 mA h g^{-1} after 50 cycles.[39] Nam *et al.*[37] prepared an electrodeposited Sn film on a porous Cu foam, revealing the formation of a Sn-Cu_6Sn_5 mixed-phase structure, obtaining an initial charge capacity of 821.5 mA h g^{-1} at 100 mA g^{-1} and retaining 58.0% of capacity after 50 cycles. The superior performance when compared to Sn electrodeposited on top of a Cu smooth foil (655.5 mA h g^{-1}) was explained by two hypotheses. On the one hand, Cu foam was better at accommodating the volumetric expansion of the active material and effectively prevented the delamination of the anode (see Figure 6.9(a) for cycling performance comparison). On the other hand, the continuous transformation of Sn into Cu_6Sn_5 during cycling could also inhibit the delamination of the active material by enhancing the bond between the active material and the Cu substrate. Figure 6.9(b) demonstrates a schematic illustration of the microstructural changes occurring at the Sn-Cu_6Sn_5/Cu foam during cycling. The as-deposited electrode contained a mixture of phases on top of the Cu foam, that gradually collapsed due to several volume changes of the active materials. Eventually, there was an increase of Cu deposits on the surface and the formation of the Li–Sn alloy.

Silicon-based materials are one of the most promising anode materials in Li-ion batteries, because of their high theoretical capacity of 4200 mA h g^{-1}.[46,47] However, these electrodes demonstrate poor conductivity and severe volumetric changes (>300%) that cause cracking and disaggregation of the electrode. The employment of a Si-based material on top of a 3D matrix can potentially relieve the stress caused by volume changes and enable the continuing reliability of the electrode. Jiang *et al.*[46] and Li *et al.*[47] prepared an amorphous film of Si based on a Cu foam using a multistep electrodeposition process and radio-frequency magnetron sputtering, respectively. Both of these materials exhibited a high specific capacity and cyclability.

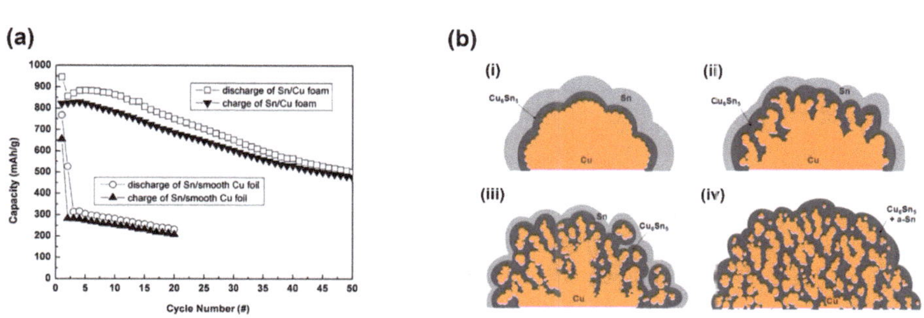

Figure 6.9 Electrochemical performances of a Sn anode electrodeposited on porous Cu foam: (a) cycling performance of the Sn-Cu_6Sn_5/Cu foam and Sn-Cu_6Sn_5/Cu smooth foil at 100 mA g^{-1}, in the potential range of 0.02 to 1.0 V (*vs.* Li/Li$^+$) and (b) schematic illustration of the phase transformation of the Sn-Cu_6Sn_5/Cu foam during cycling. Reproduced from ref. 37 with permission from Elsevier, Copyright 2012.

Recently, transition metal oxides have also been applied as active materials for anodes in Li-ion batteries, including Ti, Zn and Co. Titanium dioxide (TiO_2) was reported by An *et al.*[38] as a promising anode material for Li-ion batteries because of its safety and low volume expansion (<4%) during cycling. The incorporation of carbon-coated TiO_2 on a Cu foam enabled the enhancement of electron transport by improving the reversible specific capacity and rate capability. Wang *et al.*[42] decorated Cu foams with titanium nitride (TiN), demonstrating long-term stability and rate performances over 1900 hours of cycling as the anode for Li-ion batteries. Using Zn- and Co-based transition metal oxides, Mao *et al.*[36] developed a novel $ZnO/ZnCo_2O_4/Co_3O_4$ coating on a Cu foam (ZCO/Cu foam) using several experimental steps, including solution combustion in a glycine-nitrate system (see Chapter 1 for the combustion synthesis of foams). The structure of ZCO coating enabled the full penetration of the electrolyte, which increased the electrochemical performance of the electrode.

The use of Cu foam as a substrate for these active materials was essential to increase the energy density and allow better control of the volume-change tolerance while enhancing the electrode's energy density, compared with ZCO nanoparticles. Moreover, the structure of the coating on the Cu foam enabled the enhancement of Li ions and electron transport throughout the porous structure and improved the electrolyte contact with active material. Consequently, these phenomena contributed to a better performance of the negative electrode. The initial discharge capacity recorded for ZCO nanoparticles was 1195 mA h g^{-1} at 100 mA g^{-1}, although this valued dropped to 346 mA h g^{-1} within 20 cycles of charge/discharge. In contrast, ZCO/Cu foam demonstrated a capacity of 1389 mA h g^{-1}, maintaining the capacity of 1003 mA h g^{-1} after 35 cycles, demonstrating volume-change tolerance.[36] 3D ZCO/Cu self-supporting electrodes with a high initial discharge capacity of 1480 mA h g^{-1} were obtained, decreasing to 12 678 mA h g^{-1} after 35 cycles.

Figure 6.10(a) shows the cycling performance of these electrodes. An additional observation from the performance of ZCO-based anodes is that the capacity exhibited a gradual increase during cycling that was assigned to (1) the catalytic activity of Co and Ni elements on the structure that enhanced the pseudocapacitive behaviour; (2) the penetration of the electrolyte into inner regions that also contributed to charge storage; and (3) the synergistic effect between Zn and Co oxides.[36] Figure 6.10(b) exhibits the rate performance of the studied electrodes at various current densities from 100 mA g^{-1} to 1 A g^{-1}. In the case of the ZCO/Cu foam anode, good rate capabilities and high reversibility capacities were retained upon the change of the current density back to 100 mA g^{-1}.[36] Such behaviour is the result of negligible changes on the electrode structure.

In another work, Niu *et al.*[23] prepared a metal–organic framework (MOF) with mixed metal oxide $Cu_{0.39}Zn_{0.14}Co_{2.47}O_4$-ZnO nanoparticles that were attached to a reduced graphene oxide (rGO) and Cu foam. The presence of multi-active metal oxides on the structure enabled an enhanced

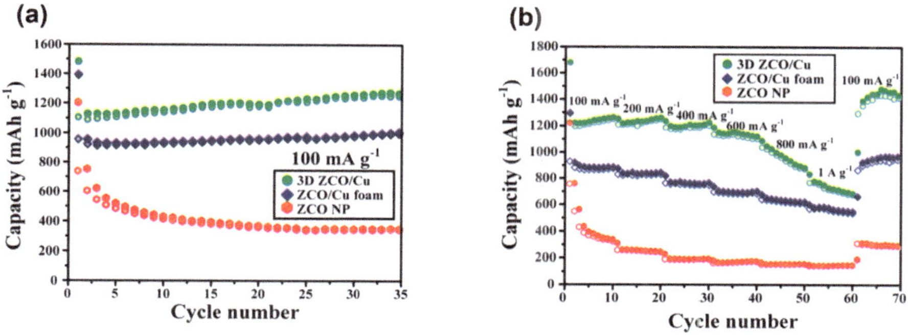

Figure 6.10 (a) Cycling performance of D ZCO/Cu, ZCO/Cu foam and ZCO NP at 100 mA g^{-1} in the range of 0.01–3.0 V (*vs.* Li/Li^+); and (b) rate performance of 3D ZCO/Cu, ZCO/Cu foam and ZCO NP at various current densities from 100 mA g^{-1} to 1 A g^{-1} in the range of 0.01–3.0 V (*vs.* Li/Li^+). Reproduced from ref. 36 with permission from Springer Nature, Copyright 2020.

capacity (initial charge and discharge capacities of 2409 and 2056 mA h g^{-1}, respectively at 0.1 A g^{-1}) and better performance. After the initial drop of capacity, the anode demonstrated long-term and reversible capacity of 1762 mA h g^{-1} with almost 100% coulombic efficiency after 500 cycles, as shown in Figure 6.11(a). Moreover, the electrode displayed good rate capability at higher applied currents, maintaining a capacity of 419 mA h g^{-1} at 10 A g^{-1} (Figure 6.11(b)).[23] The presence of Co, Cu and Zn contributes to the electrochemical activity of Li^+ ions and enables a synergistic effect on the behaviour of the anode. During cycling, the electrode undergoes an irreversible reaction between the transition metal oxides of the anode and the Li^+ ions from the electrolyte, leading to the formation of metallic particles of Co, Cu and Zn, and a Li–Zn alloy, and LiO_2 (Figure 6.11(c)). In this way, the as-formed LiO_2 is reversibly converted to Li^+, improving the reversibility and performance (eqn (6.4) is analogous to the processes described).

Other anodes made with metallic foams include materials based on alloy-composites that use commercial Cu foams as current collectors.[44,45] For instance, Liu *et al.*[44] prepared a nanoporous copper (NPCu) foam by dealloying a $Cu_{50}Al_{50}$ alloy followed by thermal oxidation that originated a composite, NPCu@CuO_2, that can be used as an anode. In the charge/discharge curves at 0.175 mA cm^{-2}, the decomposition of CuO_2 to Cu, accompanied by the formation of Li_2O, was assigned to the long plateau in Figure 6.12(a). The NPCu@CuO_2 electrode exhibited a capacity of 2.35 mA h cm^{-2} in the first discharge, due to its higher surface area; however, a drop of the capacity to 1.61 mA h cm^{-2} was noticed after the second discharge, due to the irreversible formation of the SEI layer.[44] Over ten cycles, the capacity increased slightly due to amorphization of the active material (Figure 6.12(b)).

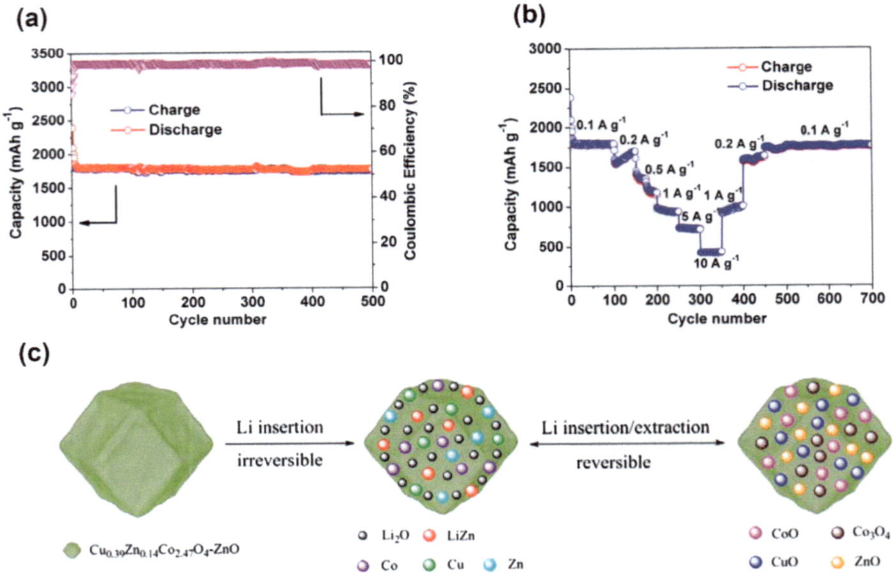

Figure 6.11 (a) Cycling performance at a current density of 0.1 A g^{-1}, in the voltage range of 0.01–3.0 V (*vs.* Li^+/Li); (b) rate capabilities at different current densities, in the voltage range of 0.01–3.0 V (*vs.* Li^+/Li); and (c) schematic illustration showing the synergistic effect between the components during the lithiation/delithiation processes. Reproduced from ref. 23 with permission from Elsevier, Copyright 2018.

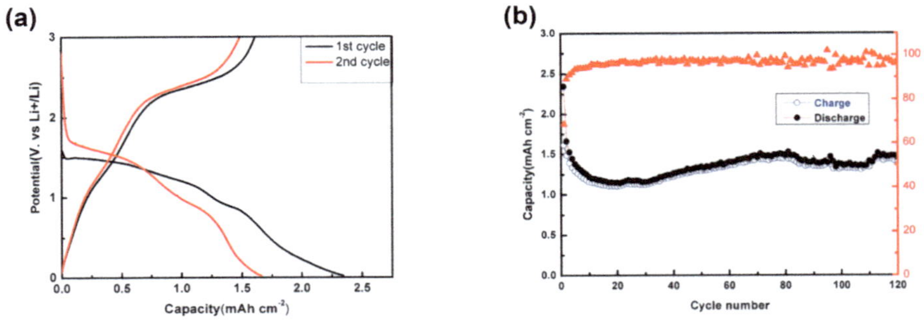

Figure 6.12 Electrochemical performance of the 3D NPCu@Cu_2O electrodes: (a) charge and discharge curves at 0.175 mA cm^{-2}, in the voltage window of 0.02–3.0 V (*vs.* Li^+/Li); and (b) galvanostatic charge and discharge cycling at a current density of 0.175 mA cm^{-2} for 120 cycles. Reproduced from ref. 44 with permission from the Royal Society of Chemistry.

In a recent work, Lu *et al.*[45] prepared a composite used as an anode material that was based on a Li–Cu alloy deposited on a Cu foam (Li–Cu/CF). The presence of lithium in the Li–Cu/CF electrode resulted in small electrode polarization and improved the cycling stability. Moreover, there was no evidence of dendrite growth upon consecutive delithiation/lithiation.

6.3.3 Alloy-based Foams

Adding multi-functional materials to the Li-ion battery electrode has been adopted as a strategy to enhance the final metrics of the device. Examples of these materials include the transition metal-based alloys. Yang *et al.*[1] used a NiCrAl alloy foam as the current collector to create a thicker electrode (300 to 600 μm) that could be loaded with a higher quantity of active material that increased notably the cell capacity. Using a cell made of metal foam having a size of 1200 μm, the device exhibited high capacity (8.8 mA h cm^{-2} at 1 mA cm^{-2}) owing to the high mass loading of the porous active material. Despite that, the results were not satisfactory enough as the charge transfer resistance also increased.[1]

Tin (Sn)-based materials have been used as anodes for Li-ion batteries despite their low theoretical capacity compared to silicon.[35] Jung *et al.*[35] prepared a Ni–Sn foam (using the electrochemical deposition method) as an anode achieving 470 mA h g^{-1} during 50 cycles. The high capacity and enhanced rate capability were related to the easier intercalation/deintercalation of Li ions from the electrolyte in the porous structure of the foam, due to the small thickness of the deposited foam.

6.3.4 Thermal Management in Li-ion Batteries

The main advantage of metal foams relies on their high surface area, attractive weight/volume features and interesting mechanical properties that can better accommodate the stress generated in the electrode materials. There are, however, some gaps to overcome, namely mechanical stability, efficient surface functionalization and optimal film thickness. Also, the pore structure and wall nano porosity and structure require in-depth characterization and evaluation to fully understand their role and maximize the performance of the batteries.

Metal foams have also been used as a key component in thermal management systems of Li-ion batteries, owing to the increase of battery temperature at higher discharge rates. If not controlled, the heat generated could be responsible for the reduction of efficiency and durability of the battery.[63,64] The thermal management of Li-ion batteries is supported by either active cooling (with air or liquid cooling medium), passive cooling or composite cooling. As part of passive cooling, Phase Change Materials (PCMs) store thermal heat in the form of sensible and latent heat, enabling uniform battery temperature.[63,65] In PCMs there is a change in the state of the material, varying from solid to liquid, or liquid to gas at a quasi-constant temperature, during heat storage.[63] However, the low thermal conductivity, high flammability and weight are the major disadvantages of PCMs in the thermal management of Li-ion batteries.[65] Metal foams have been regarded as supporting materials for PCMs, enhancing the thermal conductivity due to uniform distribution of heat by natural convection on the porous structure.[63] Mohammadian *et al.*[64] also described the influence of the permeability and porosity of aluminium metal foam in the air flow channel of the battery cell.

The insertion of aluminium metal foam, independent on the porosity, significantly improved the thermal management of the battery compared to the cell with no porous structure. These results were also described by Wang *et al.*[66]

The overview given above provides evidence of the important advantages of using different foams as current collectors or active materials. In the battery market, the use of Li-ion batteries is growing fast, as they are being used in myriads of applications that require electrochemical energy storage; however, these devices still present some limitations, which need to be overcome. Li-Ion batteries present a limited lifetime (*i.e.*, the number of charge–discharge cycles), which pose a constraint for certain long-term applications and applications with short cycle times. Also, the susceptibility of the electrolytes to degradation and decomposition raises safety concerns. There are additional concerns regarding the environmental impact of Li-ion batteries, as well as the product life cycle, recyclability and sustainability of these devices. Recycling is critical and Li-ion batteries are often criticized for different reasons. Concerns have been raised due to the ecological impact of the lithium extraction, the usage of critical raw materials and the exponential increase of their price, and many applications have been hampered. However, these drawbacks are also the driving force for the development of novel and more resilient materials to enhance the performance of Li-ion batteries, and alternative chemistries to overcome some of the intrinsic limitations of this technology. Consequently, researchers are facing an important challenge that is to find alternative and yet efficient charge storage systems based on cleaner and more abundant chemistries.

6.4 Na-ion Batteries

Presently, sodium-ion (Na-ion) batteries are emerging as a post Li-ion technology because sodium is one of the most abundant elements of the Earth's crust (Na: 23 600 *vs.* Li: 20 ppm).[67] Figure 6.13(a) compares the merits between Li, Na and K elements for rechargeable batteries, in which it is possible to observe a great opportunity for the development of cost-effective batteries using Na^+ ions. For example, sodium exhibits a reduction potential of −2.71 V similar to Li/Li^+ redox couple (−3.04 V),[68] which makes the development of these batteries possible. Moreover, in Na-ion batteries, the feasibility to work exclusively with aluminium-based current collectors, as opposed to more expensive and easier to corrode copper collectors, is being studied because no alloy is formed between sodium and aluminium.[68,69] The major feature of sodium compared to lithium is its larger ionic radius (1.02 *vs.* 0.76 Å) of Na that constrains the choice of the electrode material for Na-ion batteries with good mass transport, stability and cyclability. Figure 6.13(b) illustrates the schematics of a Na-ion battery, that work in a way similar to Li-ion batteries, in which the ion, in this case Na^+, moves forward and back between the cathode and anode during the charge and discharge of the device.

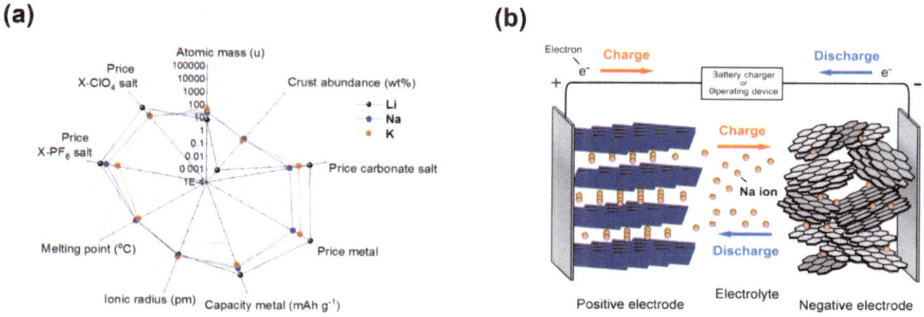

Figure 6.13 (a) Comparison of different metrics of lithium, sodium and potassium ions for rechargeable batteries. The values of the axes are plotted in logarithmic form. Reproduced from ref. 68 with permission from Elsevier, Copyright 2019. (b) Schematic illustration with a layered metal oxide as the cathode and carbon-based material as the anode. Reproduced from ref. 67 with permission from American Chemical Society, Copyright 2014.

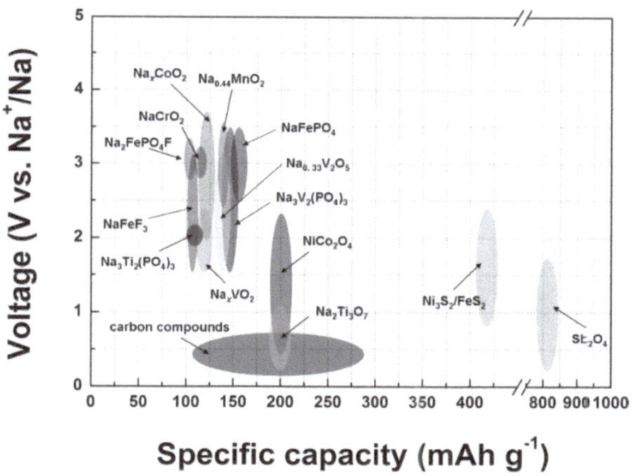

Figure 6.14 Electrode materials and corresponding electrochemical performances in Na-ion batteries. Reproduced from ref. 70 with permission from John Wiley & Sons, Copyright 2012.

Key battery performance features, such as specific capacity and operation voltage, are primarily dictated by the electrochemical properties of the electrode materials. Hence, the major challenge in Na-ion batteries depends on finding effective electrode materials. Most of the materials used as electrodes for Na-ion batteries are based on intercalating reactions, which have been used to promote the development of materials capable of inserting/de-inserting Na^+ ions. Therefore, a logical strategy to develop novel electrode materials for these batteries is to research more open structures, with adequate porosity, where Na ions can easily intercalate. Examples of electrode materials for Na-ion batteries are shown in Figure 6.14, with the corresponding specific capacity and operating voltage. For Na-ion batteries, different carbon materials (including graphite, hard/soft carbon and graphene), alloys, intercalation and conversion compounds are used as anode materials. Concomitantly, metal oxides and polyanion compounds are mainly used as cathode materials for Na-ion batteries. As it will be discussed below, several foam-supported materials have been proposed in the literature as possible anode materials, whereas fewer cathode materials supported on metal or carbon foams have been presented in the literature (although several cathode materials without foams have been already reported and summarized in review articles[67,70]). Such a lack of studies in the literature describing the use of carbon or transition metal foams as supports for Na-ion batteries represents an opportunity for the development of novel materials complying with a greenfield project for research.

6.4.1 Anode Electrode Materials

Ni foams are also used as substrates in anode materials for Na-ion batteries. Wang *et al.*[71] prepared Bi nanoflakes on Ni foam (Bi/Ni) by a replacement reaction in which Ni metal was replaced by Bi^{3+} (Figure 6.15(a)). The Bi/Ni electrode exhibited a reversible capacity of 377.1 mA h g^{-1} at 20 mA g^{-1}, and a capacity retention of 89% after 100 cycles at 200 mA g^{-1}. Chen *et al.*[72]

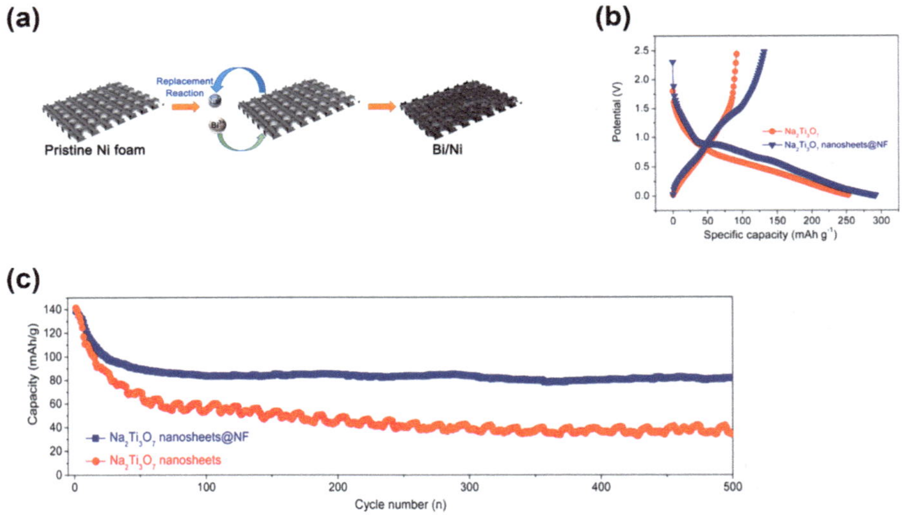

Figure 6.15 (a) Schematic diagram of the synthesis route for Bi/Ni. Reproduced from ref. 71 with permission from the Royal Society of Chemistry. (b) Charge/discharge curves of $Na_2Ti_3O_7$@NF and pure $Na_2Ti_3O_7$ at a current density of 20 mA g^{-1} and (c) cycling performance of $Na_2Ti_3O_7$@NF and pure $Na_2Ti_3O_7$ over 500 cycles at 50 mA g^{-1}. Reproduced from ref. 72 with permission from American Chemical Society, Copyright 2020.

developed layered $Na_2Ti_3O_7$ nanosheets on Ni foam ($Na_2Ti_3O_7$@NF) anode material though a hydrothermal synthesis method. The growth of the $Na_2Ti_3O_7$ nanosheets was essential to obtain an enhanced electrochemical response, as shown by the charge/discharge curves in Figure 6.15(b). The capacity of $Na_2Ti_3O_7$@NF was higher than the pure $Na_2Ti_3O_7$ without Ni foam (300.0 mA h g^{-1} *vs.* 253.3 mA h g^{-1} at 20 mA g^{-1}). This greater electrochemical performance of $Na_2Ti_3O_7$@NF was assigned to the better electrolyte penetration into the $Na_2Ti_3O_7$ nanosheets by improving the exposed active material in contact with the solution. Moreover, the cycling performance of $Na_2Ti_3O_7$@NF *vs.* pure $Na_2Ti_3O_7$ highlighted the importance of the Ni foam in the charge storage capability. As depicted in Figure 6.15(c), the pure $Na_2Ti_3O_7$ showed a loss of capacity of 75.2%, reaching 34.4 mA h g^{-1} after 500 cycles at 50 mA g^{-1}, while the decay of capacity of $Na_2Ti_3O_7$@NF was 41.3% that corresponds to 81.7 mA h g^{-1} in the same conditions.

Transition metal phosphates are very promising anode materials for Na-ion batteries owing to their high theoretical capacity (up to 1000 mA h g^{-1}), low cost and large variety.[73] Despite that, metal phosphates show problems concerning their poor electrochemical kinetics and large volume expansion that eventually lead to capacity and cycling stability fade.[73] In this way, the introduction of a conductive matrix is essential to overcome these problems. In fact, the preparation of bimetallic phosphates on top of a conductive matrix, such as Ni foam, was reported to increase the capacity and cycling stability of the electrode.

Chen *et al.*[73] synthesized a self-supported iron nickel phosphide (Fe–Ni–P) composite nanosheet on Ni foam by a hydrothermal reaction and phosphidation process. The Fe–Ni–P arrays exhibited superior performance compared to Fe_2P and Ni_2P electrodes (also deposited on top of the Ni foam), not only because of the synergy effect of the bimetallic structure, but also because of the good structural stability provided by the Ni foam. These conclusions were observed by the long-term cycling performance at 0.5 A g^{-1} and by the rate performance at different current densities, as shown in Figure 6.16(a) and (b), respectively. For instance, the Fe–Ni–P composite exhibited an initial capacity of 667.0 mA h g^{-1}, decreasing for 300.6 mA h g^{-1} after 5 cycles that then remained stable after an additional 100 cycles. On the other hand, Fe_2P and Ni_2P electrodes displayed a reversible capacity of 209.9 and 129.7 mA h g^{-1} after 100 cycles. Moreover, Fe–Ni–P delivered better rate performance than the other electrodes when the current density changes from 0.1 to 1.0 A g^{-1}.

Nickel-based selenides (NiSe) have several features that makes them promising for charge storage in Na-ion batteries. Interesting features include, for instance, large crystal layer spacing, good theoretical capacity (389 mA h g^{-1}) and abundance of raw materials for the synthesis.[74] Yu *et al.*[74] prepared NiSe/Ni foam through a one-step selenization method at different calcination temperatures (650, 750 and 850 °C for 4 hours). Figure 6.17(a) demonstrates the charge/discharge curves of NiSe/Ni foams prepared at different temperatures, in which the materials calcinated at 850 °C for 4 hours exhibited an initial capacity of 458.9 mA h g^{-1} at 0.2 mA g^{-1} and a coulombic efficiency of

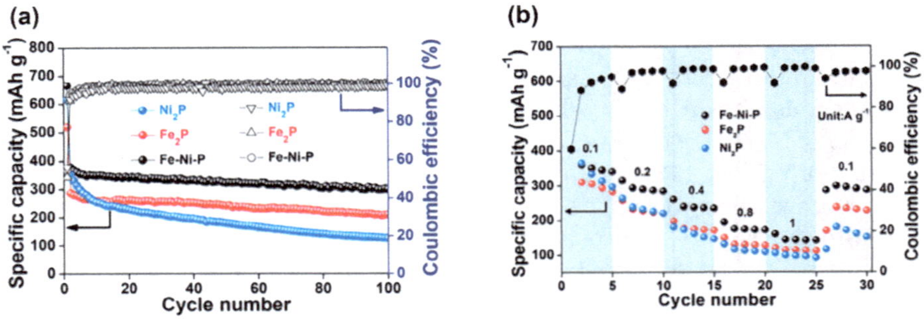

Figure 6.16 (a) Cycling performance of the Fe–Ni–P, Fe_2P and Ni_2P electrodes over 100 cycles at a current density of 0.1 A g^{-1}; and (b) rate performance of the electrode at several current densities. Reproduced from ref. 73 with permission from Elsevier, Copyright 2021.

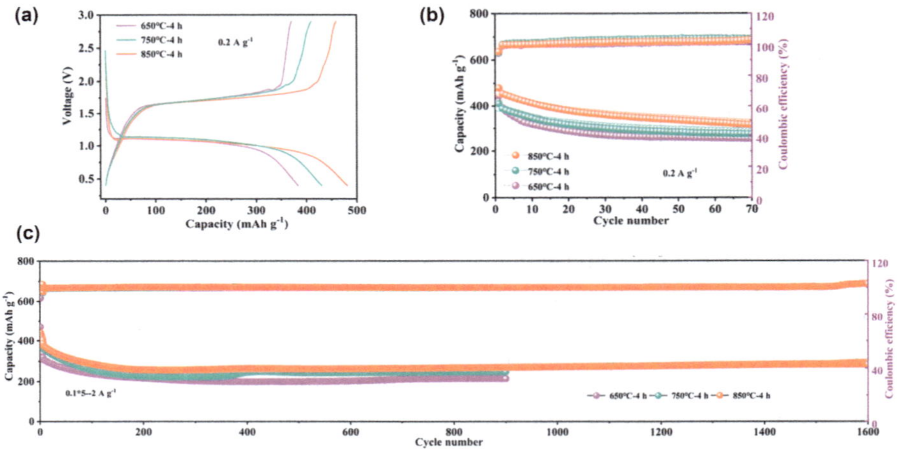

Figure 6.17 (a) Charge/discharge curves at 0.2 A g^{-1}; (b) cycling stability after 70 cycles at 0.2 A g^{-1}; and (c) long-term stability of NiSe/Ni foam prepared at 650, 750 and 850 °C for 4 hours. Reproduced from ref. 74 with permission from American Chemical Society, Copyright 2023.

95.4% that decreased to 324.2 mA h g^{-1} after 70 cycles (Figure 6.17(b)). Moreover, the NiSe/Ni foam exhibited a superior long-term stability, achieving a capacity of 219.7 mA h g^{-1} at 2 A g^{-1} after 1600 cycles (Figure 6.17(c)).

Additionally, TMO-decorated carbon foams are also used as anode materials. In contrast to Li-ion batteries where high capacity (>360 mA h g^{-1}) can be achieved, the structure of graphite materials limits the intercalation/deintercalation of Na^+ ions, reaching more modest capacities ~130 mA h g^{-1}.[70] On the other hand, hard/soft carbon composites have been used owing to their mixed properties of non- and graphitizable carbon that results in an amorphous structure with voids and interlayer regions capable of storing

Na^+ ions. Zeng *et al.*[75] have recently reported a one-step method to achieve a hard/soft hollow carbon foam by using a nickel-templated blowing strategy (denoted by HCF). The initial three cycles of charge/discharge demonstrate mechanisms with a high capacitive contribution (~89.4%) for the Na^+ storage response at 0.1 A g^{-1}, reaching a capacity of 206 mA h g^{-1}. The HCF material also demonstrated high reversibility, in which the capacity of 222 mA h g^{-1} at 0.05 A g^{-1} after cycling at 10 A g^{-1} was achieved. The doping of carbon-based foams with heteroatoms showed a significant enhancement of the electrical conductivity of the substrate that facilitated charge transfer reactions. Xu *et al.*[76] described the preparation of N-doped graphene foams (N-GF) with 6.8% of N content by annealing graphene oxide foams. The N-GF materials exhibited higher charge capacity of 1057.1 mA h g^{-1} at 0.2C and ten times higher capacity retention (13%) after 45 cycles from 0.2 to 10C compared to the graphene oxide foam without doping (809 mA h g^{-1} at 0.2C, capacity retention 1.3%). In addition, N-GF exhibited a discharge capacity retention of 69.7% and a coulombic efficiency of 98.9% after 150 cycles. Therefore, the decoration of these doped structures with WS_2 extends greatly the capacity retention over time, as reported by Wang *et al.*[77] WS_2 is a layered transition-metal dichalcogenide material that is used as an anode for Na-ion batteries, because of its large theoretical specific capacity of 432 mA h g^{-1}.[77] Being a layered structure, it facilitates the intercalation/deintercalation of Na^+ ions into the interlayer region (0.62 nm spacing), and keeps large volume expansion from occurring during charge/discharge.[77] The incorporation of N-doped graphene quantum dots-decorated WS_2 uniformly distributed on top of carbon foam (NGQDs-WS_2/3DCF) showed an initial discharge and charge capacity of 657.9 and 428.8 mA h g^{-1}, respectively; however, with low coulombic efficiency due to the formation of SEI film. The discharge curves demonstrate multiple plateaus that correspond to the insertion of Na^+ ions and conversion of WS_2 into W nanoparticles.[77] The reverse process was observed in the charge curves. The NGQDs-WS_2/3DCF material exhibited a high capacity retention at various current densities from 50 to 5000 mA g^{-1}, with an initial capacity of 460.9 mA h g^{-1} at 50 mA g^{-1}. Compared with previous studies of N-doped carbon foams, the NGQDs-WS_2/3DCF exhibited an outstanding 97.1% capacity retention after 1000 cycles assigned to the presence of WS_2 combined with N-doped graphene quantum dots.[77]

Ma *et al.*[78] described the preparation of transitional metal sulfides (TMSs) anchored on 3D N-doped carbon foam *via* coordination and hydrogen-bond adsorption. The TMSs@3DNCF structures prepared were Co_9S_8@C/3DNCF, FeS@C/3DNCF, Ni_3S_2@C/3DNCF, MnS@C/3DNCF and ZnS@C/3DNCF. All of these structures revealed high specific capacities, superior rate capability and longer cycle durability as self-standing anodes for high-performance Na-ion batteries.

Chen *et al.*[79] prepared a Se-doped chalcogen-based material using Bi_2S_3 supported on carbon foam ($Bi_2S_{3-x}Se_x$/CF) by a three-step procedure of pyrolyzing-hydrothermal-heating. Bi_2S_3 exhibits a high specific theoretical capacity of 625 mA h g^{-1}; however, volume changes caused by the continuous cycling with eventual structural failure of the electrode restrain the long-term stability. The Se-doping of Bi_2S_3 was the strategy that Chen *et al.*[79] reported, in which the electronic conductivity of the hybrid material increased as well as the mechanical stability and electrochemical performance. For instance, the hybrid material exhibited a high specific capacity of 441.8 mA h g^{-1} at 100 mA g^{-1} after 100 cycles with coulombic efficiency of ~100%.

6.4.2 Cathode Electrode Materials

Sodium-containing layered oxide cathode materials (*i.e.*, Na_xMO_2, $x < 1$, M = Co, Ni, Fe, *etc.*) were initially used as cathode materials for the intercalation and deintercalation of Na^+ ions.[80] In the case of Na_xMO_2 materials, both O3 and P2 crystalline structures are present on the charge/discharge processes, in which Na^+ ions are intercalated or deintercalated. Figure 6.18 shows a comparison of the electrochemical behaviour of a Li/$LiCoO_2$ cell in Li-ion batteries with a Na/$NaCoO_2$ cell in Na-ion batteries, highlighting the larger operating voltage of $LiCoO_2$ compared with $NaCoO_2$. This lower voltage range entails a lower available energy density of Na-ion systems.[67] With this example, the authors proposed that the distinctive chemistry from the Li system needs to overcome the problem of low energy density for Na-ion batteries. Interestingly, it is possible to observe multiple voltage plateaus, which reflects different phase transitions occurring at the cathode material.

Taking advantage of 3D and highly porous Ni foam, Gao *et al.*[81] prepared $Na_{0.67}CoO_2$ arrays on top of the foam using two-step reactions. The cycling stability was analysed at 1.4C (equivalent to 189 mA g^{-1}), resulting in an initial capacity of 107.5 mA h g^{-1} that slowly increased to 128.1 mA h g^{-1} after 40 cycles.[81] The increase of capacity was assigned to the activation of the electrode by the penetration of the electrolyte into the interconnected framework. After 100 cycles, the discharge capacity remained at 123.6 mA h g^{-1} with a high coulombic efficiency value (99.5%), demonstrating high reversibility (Figure 6.19(a)). Moreover, with the progressive variation of the

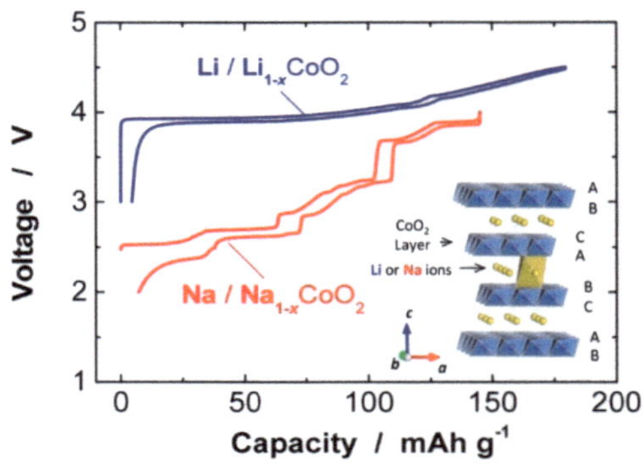

Figure 6.18 Charge/discharge curves of Li/$LiCoO_2$ and Na/$NaCoO_2$ cells. Inset figure shows the layered metal oxide structure. Reproduced from ref. 67 with permission from American Chemical Society, Copyright 2014.

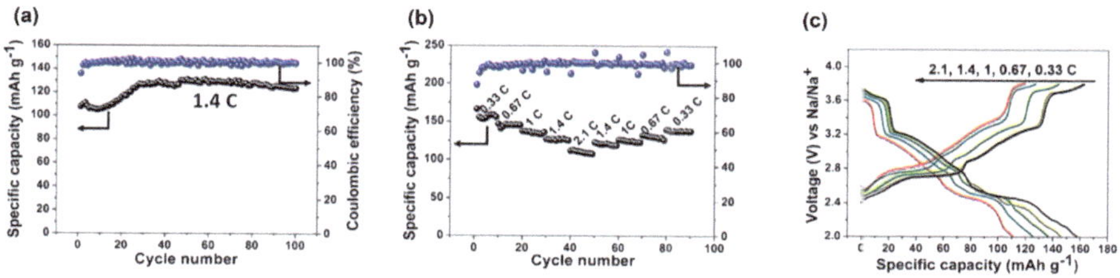

Figure 6.19 Cycling performance of $Na_{0.67}CoO_2$ on Ni foam: (a) gravimetric capacity and coulombic efficiency at 1.4C for 100 cycles; (b) rate performance at several current rates; and (c) galvanostatic charge and discharge cycling at different current rates. Reproduced from ref. 81 with permission from John Wiley & Sons, Copyright 2019.

current rate, the $Na_{0.67}CoO_2$-based cathode exhibited high reversible capacity at 0.33C (Figure 6.19(b)). The multiple plateaus observed in Figure 6.19(c) are assigned to the existence of biphasic domains during the insertion/extraction process of Na^+ ions.

Other foam-based current collectors for cathode electrodes described in the literature are based on carbon foams.[79,82–84] For instance, Li *et al.*[82] prepared two polyanion-based materials: 3D hybrid foam of $Na_7V_4(P_2O_7)_4(PO_4)$ and $Na_7V_3(P_2O_7)_4$ deposited on biomass-derived porous carbon. The porous structure of the hybrid materials enabled an easier intercalation/deintercalation of sodium. Moreover, these two vanadium-based materials have a high operating potential window of ~4.0 V *vs.* Na/Na^+, being therefore promising cathode materials for Na-ion batteries. The application of the polyanion materials on top of the carbon foam was essential for increasing the poor conductivity, and inferior stability of these materials.[82] $Na_7V_4(P_2O_7)_4(PO_4)$ and $Na_7V_3(P_2O_7)_4$ demonstrated a high capacity retention of 94 and 91% respectively, after 800 cycles alternating 20C and 3C applied current.

6.5 Other Batteries

The quest for more advanced electrode materials has led to many new materials being explored that have been reported in various studies, some of which address the use of metal foams on other types of batteries. Ni–metal hydride batteries were state-of-the-art for automotive applications that, in certain cases, relied on Ni foams as current collectors. These batteries have higher energy density compared to Pb–acid batteries, but suffer some drawbacks, such as memory effect and limited lifetime cycling performance. The rising price of Ni and its limited lifetime (typically below 1000 cycles) do not favour the market of Ni batteries and the dominance of Li-ion batteries has been notorious throughout the 21st century. Nonetheless, nickel foams were a milestone in the development of nickel metal-hydride batteries.[85–88] Besides Ni–metal batteries, Ni foams are a promising material for the next-generation of nickel–zinc (Ni–Zn),[89–91] Zn–air[92,93] and alkaline Ni–Fe[94] batteries.

As evidenced by the current literature review presented in this chapter, metal-based foam materials contribute to improve the performance of the different battery chemistries, However, their massive implementation is still being challenged by the lack of some fundamental understanding, for example, regarding the effect of the properties of these materials in the SEI formation in Li-ion batteries; an issue that requires more research. Additionally, the need for large-scale manufacturing methods that can cope with the current battery demand, and the cost-effectiveness of using metallic foams as a replacement for the well-known foils used as current collectors, whose market is very well established, have to be addressed. The mixture of these scientific and market uncertainties still poses a big challenge to the mass-adoption of metallic foams in commercial batteries. Nonetheless, the need for more competitive performance and the development of different chemistries may drive future development and implementation of metallic foams or their composites with different materials including different forms of carbon. Certainly, these materials present very promising features that can assist in overcoming some of the technological problems that we are facing today.

References

1. G.-F. Yang, K.-Y. Song and S.-K. Joo, *RSC Adv.*, 2015, **5**, 16702–16706.
2. K. Ji, C. Xu, H. Zhao and Z. Dai, *J. Power Sources*, 2014, **248**, 307–316.
3. Z. Yang, J. Zhang, M. C. Kintner-Meyer, X. Lu, D. Choi, J. P. Lemmon and J. Liu, *Chem. Rev.*, 2011, **111**, 3577–3613.
4. E. Gyenge, J. Jung and A. A. Snaper, *US Pat.*, US20060292448A1, 2006.
5. K. C. Kelley and J. J. Votoupal, *US Pat.*, US6979513B2, 2005.
6. K. C. Kelley and M. J. Maroon, *US Pat.*, US8399134B2, 2013.
7. C. C. Kelly, M. J. Maroon and C. F. Ostermeyer, *Japan Pat.*, JP5005886B2, 2012.
8. L. Ma, Z. Nie, X. Xi, B. Chen and Y. Chen, *J. Porous Mater.*, 2012, **20**, 557–562.
9. Y.-I. Jang, N. J. Dudney, T. N. Tiegs and J. W. Klett, *J. Power Sources*, 2006, **161**, 1392–1399.
10. G. Costanza and M. E. Tata, in *REWAS 2013*, Springer, 2013, 272–278.
11. A. Irretier and J. Banhart, *Acta Mater.*, 2005, **53**, 4903–4917.
12. C.-S. Dai, B. Zhang, D.-L. Wang, T.-f. Yi and X.-G. Hu, *J. Alloys Compd.*, 2006, **422**, 332–337.
13. S. M. Tabaatabaai, M. S. Rahmanifar, S. A. Mousavi, S. Shekofteh, J. Khonsari, A. Oweisi, M. Hejabi, H. Tabrizi, S. Shirzadi and B. Cheraghi, *J. Power Sources*, 2006, **158**, 879–884.
14. C. Dai, T. Yi, D. Wang and X. Hu, *J. Power Sources*, 2006, **158**, 885–890.
15. C. S. Dai, B. Zhang, D. L. Wang, T. F. Yi and X. G. Hu, *Mater. Chem. Phys.*, 2006, **99**, 431–436.
16. K. Wróbel, *Int. J. Electrochem. Sci.*, 2016, 8927–8937.
17. N. Rees and R. Fuller, The Toxic Truth: Children's Exposure to Lead Pollution Undermines a Generation of Future Potential, https://www.unicef.org/reports/toxic-truth-childrens-exposure-to-lead-pollution-2020, accessed May 2023.

18. B. Dunn, H. Kamath and J.-M. Tarascon, *Science*, 2011, **334**, 928–935.
19. Y. Chen, Y. Kang, Y. Zhao, L. Wang, J. Liu, Y. Li, Z. Liang, X. He, X. Li, N. Tavajohi and B. Li, *J. Energy Chem.*, 2021, **59**, 83–99.
20. J. B. Goodenough and K. S. Park, *J. Am. Chem. Soc.*, 2013, **135**, 1167–1176.
21. N. Nitta, F. Wu, J. T. Lee and G. Yushin, *Mater. Today*, 2015, **18**, 252–264.
22. N. A. Kaskhedikar and J. Maier, *Adv. Mater.*, 2009, **21**, 2664–2680.
23. J.-L. Niu, H.-J. Peng, C.-H. Zeng, X.-M. Lin, P. Sathishkumar, Y.-P. Cai and A.-W. Xu, *Chem. Eng. J.*, 2018, **336**, 510–517.
24. D. J. Noelle, M. Wang and Y. Qiao, *J. Power Sources*, 2018, **399**, 125–132.
25. Y. Yu, C. H. Chen, J. L. Shui and S. Xie, *Angew Chem. Int. Ed. Engl.*, 2005, **44**, 7085–7089.
26. J. H. Um, H. Park, M. Choi, H. Choi, Y.-H. Cho, Y.-E. Sung and H. Choe, *US Pat.*, US10265773B2, 2019.
27. Z. Zhang, J. Wang, X. Yan, S. Zhang, W. Yang, Z. Zhuang and W.-Q. Han, *Energy Storage Mater.*, 2020, **29**, 332–340.
28. G. Huang, P. Lou, G.-H. Xu, X. Zhang, J. Liang, H. Liu, C. Liu, S. Tang, Y.-C. Cao and S. Cheng, *J. Alloys Compd.*, 2020, **817**, 152753.
29. Q. Zhang, J. Wang, J. Dong, F. Ding, X. Li, B. Zhang, S. Yang and K. Zhang, *Nano Energy*, 2015, **13**, 77–91.
30. J. Zhang, R. Chu, Y. Chen, H. Jiang, Y. Zhang, N. M. Huang and H. Guo, *Ceram. Int.*, 2018, **44**, 16219–16226.
31. N. Feng, D. Hu, P. Wang, X. Sun, X. Li and D. He, *Phys. Chem. Chem. Phys.*, 2013, **15**, 9924–9930.
32. M. Fesahat, H. Vafaeenezhad, M. H. Esnaashary, A. Yazdi, F. Hosseinabadi and M. Soltanieh, *Mater. Sci. Eng., A*, 2021, **825**, 141895.
33. K. Y. Song, G. S. Jang, J. Tao, J. H. Lee and S. K. Joo, *J. Electrochem. Soc.*, 2016, **163**, A2981–A2987.
34. M. A. Rahman, M. M. Rahman and G. Song, *Energy Storage*, 2021, **4**, e278.
35. H. R. Jung, E. J. Kim, Y. J. Park and H. C. Shin, *J. Power Sources*, 2011, **196**, 5122–5127.
36. H. Mao, P. Shen, G. Yang, L. Zhao, X. Qiu, H. Wang and Q. Jiang, *J. Mater. Sci.*, 2020, **55**, 11462–11476.
37. D. H. Nam, R. H. Kim, D. W. Han and H. S. Kwon, *Electrochim. Acta*, 2012, **66**, 126–132.
38. C.-s. An, B. Zhang, L.-b. Tang, B. Xiao, Z.-j. He and J.-c. Zheng, *Ceram. Int.*, 2019, **45**, 13144–13149.
39. R. H. Kim, D. W. Han, D. H. Nam, J. Kim and H. S. Kwon, *J. Electrochem. Soc.*, 2010, **157**, D269–D273.
40. Q. Li, S. Hu, H. Wang, F. Wang, X. Zhong and X. Wang, *Electrochim. Acta*, 2009, **54**, 5884–5888.
41. A. Etiemble, J. Adrien, E. Maire, H. Idrissi, D. Reyter and L. Roué, *Mater. Sci. Eng., B*, 2014, **187**, 1–8.
42. Y. Wang, W. Q. Zhang, Y. Y. Qi, S. F. Wang, P. Liu, X. Y. Wei, Y. H. Yu, W. W. Sun, X. Z. Zhao and Y. M. Liu, *J. Alloys Compd.*, 2021, **874**, 159916.
43. M. Vainoris, H. Cesiulis and N. Tsyntsaru, *Coatings*, 2020, **10**, 822.

44. D. Liu, Z. Yang, P. Wang, F. Li, D. Wang and D. He, *Nanoscale*, 2013, **5**, 1917–1921.
45. Z. Lu, Z. Tai, Z. Yu, A. P. LaGrow, O. Bondarchuk, J. P. S. Sousa, L. Meng, Z. Peng and L. Liu, *Mater. Today Energy*, 2021, **22**, 100871.
46. T. Jiang, S. Zhang, X. Qiu, W. Zhu and L. Chen, *Electrochem. Commun.*, 2007, **9**, 930–934.
47. H. Li, F. Cheng, Z. Zhu, H. Bai, Z. Tao and J. Chen, *J. Alloys Compd.*, 2011, **509**, 2919–2923.
48. H. Wang, Q. Pan, X. Wang, G. Yin and J. Zhao, *J. Appl. Electrochem.*, 2009, **39**, 1597–1602.
49. C. Wang, D. Wang, Q. Wang and H. Chen, *J. Power Sources*, 2010, **195**, 7432–7437.
50. S. Ni, T. Li, X. Lv, X. Yang and L. Zhang, *Electrochim. Acta*, 2013, **91**, 267–274.
51. Y. Song, J. Hwang, S. Lee, B. Thirumalraj, J.-H. Kim, P. Jenei, J. Gubicza and H. Choe, *Adv. Eng. Mater.*, 2020, **22**, 200351.
52. C. Liu, C. Li, K. Ahmed, Z. Mutlu, C. S. Ozkan and M. Ozkan, *Sci. Rep.*, 2016, **6**, 29183.
53. W. Yang, G. Cheng, C. Dong, Q. Bai, X. Chen, Z. Peng and Z. Zhang, *J. Mater. Chem. A*, 2014, **2**, 20022–20029.
54. X. H. Huang, J. P. Tu, X. H. Xia, X. L. Wang and J. Y. Xiang, *Electrochem. Commun.*, 2008, **10**, 1288–1290.
55. Y. Fan, X. He, H. Li, Y. Huang, C. Sun, H. Liu, E. Huangzhang, F. Sun, X. Zhao and J. Nan, *Chem. Eng. J.*, 2022, **450**, 138384.
56. S. Liu, H. Zhang, X. Liu, Y. Yang, C. Chi, S. Wang, J. Xue, T. Hao, J. Zhao and Y. Li, *J. Energy Chem.*, 2021, **58**, 124–132.
57. H. Ruan, Y. Li, H. Qiu and M. Wei, *J. Alloys Compd.*, 2014, **588**, 357–360.
58. D. Li, X. Li, X. Hou, X. Sun, B. Liu and D. He, *Chem. Commun.*, 2014, **50**, 9361–9364.
59. Q. Wang, R. Gao and J. Li, *Appl. Phys. Lett.*, 2007, **90**, 143107.
60. P. Poizot, S. Laruelle, S. Grugeon, L. Dupont and J. M. Tarascon, *Nature*, 2000, **407**, 496–499.
61. K. Mizushima, P. C. Jones, P. J. Wiseman and J. B. Goodenough, *Mater. Res. Bull.*, 1980, **15**, 783–789.
62. X. Rui, H. Tan and Q. Yan, *Nanoscale*, 2014, **6**, 9889–9924.
63. B. Buonomo, D. Ercole, O. Manca and F. Menale, *Energy Procedia*, 2018, **148**, 1175–1182.
64. S. K. Mohammadian, S. M. Rassoulinejad-Mousavi and Y. Zhang, *J. Power Sources*, 2015, **296**, 305–313.
65. H. Shi, M. Cheng, Y. Feng, C. Qiu, C. Song, N. Yuan, C. Kang, K. Yang, J. Yuan and Y. Li, *Energies*, 2023, **16**, 876.
66. Y. Wang, Y. Yu, Z. Jing, C. Wang, G. Zhou and W. Zhao, *Int. J. Heat Mass Transfer*, 2021, **167**, 120827.
67. N. Yabuuchi, K. Kubota, M. Dahbi and S. Komaba, *Chem. Rev.*, 2014, **114**, 11636–11682.
68. W. Zhang, F. Zhang, F. Ming and H. N. Alshareef, *EnergyChem*, 2019, **1**, 100012.
69. A. Bauer, J. Song, S. Vail, W. Pan, J. Barker and Y. Lu, *Adv. Energy Mater.*, 2018, **8**, 1702869.

70. S.-W. Kim, D.-H. Seo, X. Ma, G. Ceder and K. Kang, *Adv. Energy Mater.*, 2012, **2**, 710–721.
71. L. Wang, C. Wang, F. Li, F. Cheng and J. Chen, *Chem. Commun.*, 2017, **54**, 38–41.
72. Z. Chen, Q. Zhang, L. Lu, X. Chen, S. Wang, C. Xin, B. Xing and C. Zhang, *Energy Fuels*, 2020, **34**, 3901–3908.
73. G. Chen, L. Gao, L. Zhang and X. Yang, *J. Electroanal. Chem.*, 2021, **895**, 115420.
74. L. Yu, L. Shao, R. Pan, J. Guan, J. Lin, X. Shi, J. Cai, J. Sun, C. Chen, Y. Wu and Z. Sun, *ACS Sustainable Chem. Eng.*, 2023, **11**, 10324–10332.
75. J. Zeng, T. Wang, X. Gu, H. Zhu, C. Xu, D. Sun, C. Ge, R. Ding, J. Li, J. Liu, J. Rong, X. Wang and X. Jiang, *FlatChem*, 2023, **40**, 100508.
76. J. Xu, M. Wang, N. P. Wickramaratne, M. Jaroniec, S. Dou and L. Dai, *Adv. Mater.*, 2015, **27**, 2042–2048.
77. Y. Wang, D. Kong, S. Huang, Y. Shi, M. Ding, Y. Von Lim, T. Xu, F. Chen, X. Li and H. Y. Yang, *J. Mater. Chem. A*, 2018, **6**, 10813–10824.
78. X. Ma, X. Xiong, P. Zou, W. Liu, F. Wang, L. Liang, Y. Liu, C. Yuan and Z. Lin, *Small*, 2019, **15**, e1903259.
79. X. Chen, Y. Hong, X. Ge, C. Li, X. Miao, P. Wang, Z. Zhang and L. Yin, *J. Alloys Compd.*, 2020, **825**, 153901.
80. C. Delmas, J.-J. Braconnier, C. Fouassier and P. Hagenmuller, *Solid State Ionics*, 1981, **3/4**, 165–169.
81. L. Gao, S. Chen, L. Zhang and X. Yang, *ChemElectroChem*, 2019, **6**, 947–952.
82. Q. Li, B. Lin, S. Zhang and C. Deng, *J. Mater. Chem. A*, 2016, **4**, 5719–5729.
83. Y. Shao, J. Xiao, W. Wang, M. Engelhard, X. Chen, Z. Nie, M. Gu, L. V. Saraf, G. Exarhos, J. G. Zhang and J. Liu, *Nano Lett.*, 2013, **13**, 3909–3914.
84. Y. Chen, J. Li, Y. Lai, M. Yin and Z. Zhang, *J. Electroanal. Chem.*, 2018, **810**, 207–215.
85. M. Yao, K. Okun, T. Iwaki, M. Kato, K. Harada, J.-J. Park, S. Tanase and T. Sakai, *J. Electrochem. Soc.*, 2007, **157**, A709–A714.
86. Y. Wang, D. Cao, G. Wang, S. Wang, J. Wen and J. Yin, *Electrochim. Acta*, 2011, **56**, 8285–8290.
87. W. Metzger, R. Westfall, A. Hermann and P. Lyman, *Int. J. Hydrogen Energy*, 1998, **23**, 1025–1029.
88. W. Metzger, R. Westfall and A. Hermann, *Int. J. Hydrogen Energy*, 1998, **23**, 1179–1183.
89. Y. Huang, M. Li, S. Chen, P. Sun, X. Lv, B. Li, L. Fang and X. Sun, *Appl. Surf. Sci.*, 2021, **562**, 150222.
90. Z. Lu, X. Wu, X. Lei, Y. Li and X. Sun, *Inorg. Chem. Front.*, 2015, **2**, 184–187.
91. Y. Xia, D. Zhu, S. Si, D. Li and S. Wu, *J. Power Sources*, 2015, **283**, 125–131.
92. P. Tan, B. Chen, H. Xu, W. Cai, W. He and M. Ni, *Appl. Catal., B*, 2019, **241**, 104–112.
93. C. Li, X. Shi, S. Liang, X. Ma, M. Han, X. Wu and J. Zhou, *Chem. Eng. J.*, 2020, **379**, 122248.
94. X. Li and D. Xiao, *ChemElectroChem*, 2020, **7**, 3098–3105.

CHAPTER 7

Energy-related Electrocatalysis Applications

7.1 Introduction

The depletion of fossil fuels and the continuous growing threat of environmental pollution have caused a global energy crisis, stimulating a remarkable effort for the development of novel renewable energy storage and conversion technologies that rely on sustainable chemical methods. One of the most promising energy conversion systems is based on the hydrogen cycle, which includes several electrochemical processes.[1] First, the water splitting reaction generates gaseous molecular hydrogen and oxygen in an electrolytic cell, with the hydrogen evolution reaction (HER) and the oxygen evolution reaction (OER) occurring simultaneously at the cathode and the anode, respectively.[2,3] Subsequently, the produced hydrogen can be oxidized on a hydrogen–oxygen fuel cell: the hydrogen oxidation reaction (HOR) takes place at the anode and the oxygen reduction reaction (ORR) occurs at the cathode. The hydrogen evolution and oxidation reactions proceed at high rates, while electrochemical oxygen reactions have slower kinetics and involve multiple proton-coupled electron transfer steps.[3] In Figure 7.1 it is possible to observe a representation of the half-cell reaction equations and typical steady-state polarization curves for the aforementioned processes. The four reactions can be categorized into two reversible reaction couples, one being the hydrogen-involving HER and HOR with the equilibrium potential of 0 V *vs.* the reversible hydrogen electrode (RHE) and the other being the oxygen-involving ORR and OER with the equilibrium potential of 1.23 *vs.* RHE.

Transition Metal-based Nanofoams for Electrochemical Systems: Manufacturing, Characterization and Applications
By Maria de Fátima Montemor, Diana M. Fernandes, Alberto Adan-Mas, Ana Catarina Alves, Gabriel Garcia Carvalho and Inês S. Marques

Published by the Royal Society of Chemistry, www.rsc.org

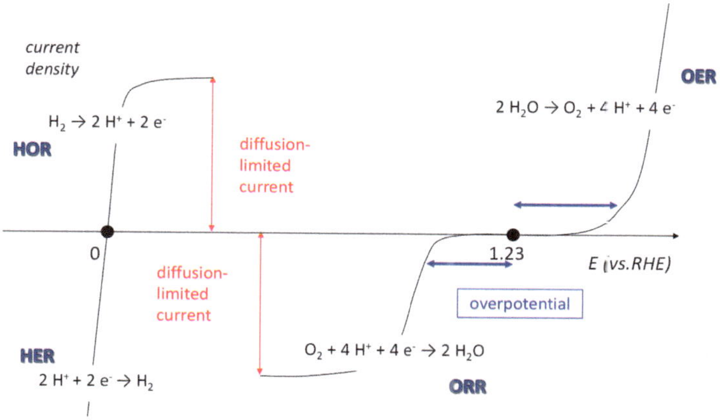

Figure 7.1 LSV curves for the two reversible reaction couples and their overall reaction equations. The lines are not drawn to scale.

Up to now, platinum-based materials are the most effective electrocatalysts for the ORR and HER, while RuO_2 and IrO_2 are the best for the OER. However, their high cost, low availability and poor durability have stimulated the development of novel electrocatalysts over the past years.[2] Among the potential alternative materials, we find 3D foam-based electrocatalysts. Their intrinsic properties like large pore volume, high specific surface area, enhanced electrolyte penetration and porous network structure are extremely beneficial to improve mass transport of reactants to the electrode surface and to pave 3D pathways for electron transfer. Moreover, foams can either be used as electrode materials by themselves or as a support for other electroactive species like carbon-based materials and transition or noble metal-based species.

7.2 Fundamentals of the HER, OER, and ORR

7.2.1 Hydrogen Evolution Reaction

Hydrogen is one of the cleanest and most promising energy carriers. The electrolysis of water, involving the hydrogen and oxygen evolution reactions, is an efficient and environmentally friendly way to produce hydrogen and oxygen. The hydrogen evolution reaction is thought to involve three possible reaction steps in acidic conditions.[4] When an electron is transferred to the cathode, it will capture a proton in the electrolyte to produce an intermediate state of an adsorbed hydrogen atom (H_{ad}) on top of the electrode surface (Volmer step). Then, if the H_{ad} coverage is relatively low, the H_{ad} prefers to couple with another electron and catch a new proton to generate H_2 (Heyrovsky step) or two H_{ad} can combine, which increases the H_{ad} coverage, giving rise to H_2 (Tafel step). Regardless of the pathway, the initial process of HER always involves the adsorption of a proton. So, the free energy of hydrogen adsorption (ΔG_H) is generally used to theoretically evaluate the activity of HER electrocatalysts.[5] This means that the lower the absolute value of ΔG_H, the faster the overall turnover rate. For this reason, platinum is the most active single-metal catalyst known for HER ($\Delta G_H \approx$ zero).

$$H^+(aq) + e^- \rightarrow H_{ad} \quad \text{(Volmer step)}$$

$$H_{ad} + H^+(aq) + e^- \rightarrow H_2(g) \quad \text{(Heyrovsky step)}$$

$$H_{ad} + H_{ad} \rightarrow H_2(g) \quad \text{(Tafel step)}$$

In alkaline conditions, the HER also follows three possible steps.[6] Initially, a water molecule couples with an electron producing H_{ad} on the cathode surface (Volmer step). At first, the H_{ad} tends to couple with another electron and grab a new water molecule to produce H_2 (Heyrovsky step). Then, when the amount of H_{ad} increases, the adjacent H_{ad} begin to combine with each other producing H_2 (Tafel step).

$$H_2O + e^- \rightarrow OH^- + H_{ad} \quad \text{(Volmer step)}$$

$$H_{ad} + H_2O + e^- \rightarrow OH^- + H_2(g) \quad \text{(Heyrovsky step)}$$

$$H_{ad} + H_{ad} \rightarrow H_2(g) \quad \text{(Tafel step)}$$

To evaluate the electrocatalytic activity of a particular HER electrocatalyst and compare the performances between selected materials, some important parameters must be carefully measured/calculated. These include the overpotential (η), apparent total electrode activity, Tafel slope, stability, faradic efficiency and turnover frequency.[3,7]

- *Overpotential*

The overpotential (η) is one of the most important parameters to evaluate the performance of selected electrocatalysts (Figure 7.2(a)).[8] Ideally, the applied

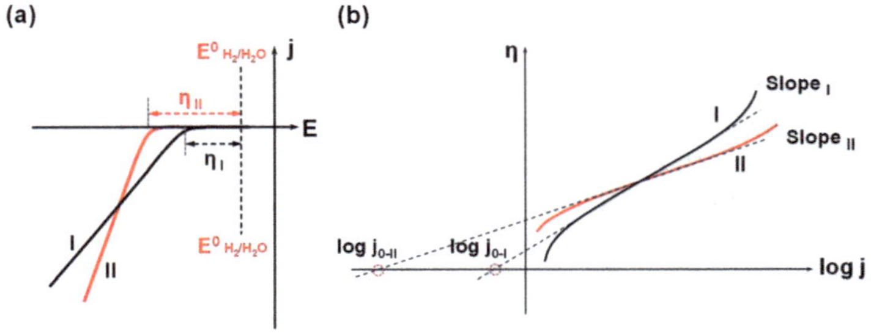

Figure 7.2 Schematic HER polarization curves on two different electrocatalysts with their onset overpotentials indicated (a) and schematic Tafel plots on two different electrocatalysts with their Tafel slopes and exchange current densities indicated (b). Reproduced from ref. 8 with permission from the Royal Society of Chemistry.

potential for driving a specific reaction and the reaction potential at equilibrium should be the same. However, this is not the case in real systems where the applied potential is commonly much higher than that at equilibrium in order to overcome the electrode kinetic barrier of the reaction. According to the Nernst equation, the applied potential can be expressed as eqn (7.1), where E is the applied potential, $E°$ is the formal potential of the overall reaction, T is the absolute temperature, R is the universal gas constant, F is Faraday constant, n is the number of transferred electrons in the reaction, and C_O and C_R are the concentrations of oxidized and reduced reagents, respectively. The overpotential (η) is the difference between the electrode and standard potentials ($\eta = E - E_0$). In general, the overpotential (η_{10}) is commonly referred to as a value that has to be applied to achieve a specified current density of $j = 10$ mA cm^{-2}, which is the current density required for a 12.3% efficiency in a solar-to-hydrogen device during solar-powered water splitting, which would be required to produce a cost-competitive photoelectrochemical water splitting system.[7] It is commonly accepted that a lower overpotential of an electrocatalyst indicates its superior electrocatalytic ability for the selected reaction.

$$E = E^{\circ} + \frac{RT}{nF}\ln\frac{C_O}{C_R} \tag{7.1}$$

- *Total electrode activity*

The total electrode activity is usually estimated by performing cyclic voltammetry (CV) or linear sweep voltammetry (LSV). However, since non-faradaic capacitive currents can also contribute to the total observed current, conducting more precise tests is typically necessary. These involve the measurement of the material steady-state currents at different voltages. Typically, the electrode geometric area is used to normalize the currents obtained and this is defined as the current density, j, and is expressed in mA $cm_{geometric}^{-2}$.[5] The intrinsic electrode activity can also be determined, normalizing the currents obtained by the electrocatalytic active surface area (ECSA), which reflects the total number of active sites on a given surface for a certain reaction. To compare the activities between samples, researchers usually resort to the overpotential (η_{10}).[7]

- *Tafel slope (TS)*

The Tafel plot represents the dependence of steady-state current densities at different overpotentials, and is obtained by the fit of the linear part of the Tafel plot (see Figure 7.2(b)) by applying the Tafel equation:

$$\eta = a + b\log j \quad \text{(Tafel equation)}$$

where η defines the overpotential, j denotes the current density, and b is the Tafel slope.[9] This parameter is generally used to infer the electrocatalytic mechanism of the electrode reaction. For values close to 30 mV dec^{-1}, the Tafel step is the rate-determining step, while for values closer to 40 mV dec^{-1} and 120 mV dec^{-1} the Heyrovsky and Volmer steps become the rate-determining steps, respectively.[9–11] Considering that the HER involves both the adsorption and release of the hydrogen atoms on the electrocatalyst surface, a good electrocatalyst should have a good balance between these two steps. Another important physical parameter that can be determined from the Tafel analysis is the exchange current density (j_0), which can be obtained for $\eta = 0$ and relates to the intrinsic electrocatalytic activity of the ECs under equilibrium conditions.[3] Therefore, lower Tafel slopes and higher j_0 are desirable, especially for high current applications such as in water electrolyzers.

- *Faradaic efficiency*

In a more general way, the faradaic efficiency is the efficiency of electron participation in a specific electrochemical reaction. Regarding HER, faradaic efficiency is defined as the ratio between the amount of H_2 experimentally measured and the theoretical value, the latter being calculated based on a 100% faradaic yield.[7]

- *Turnover frequency (TOF)*

The turnover frequency is the number of chemical transformations that can be driven by an electrocatalytic site per unit of time.[5] However, TOF values can be imprecise as a result of the heterogeneous nature of most current electrocatalysts. Catalytic sites may consist on both accessible and inaccessible catalytic groups or surface atoms/internal atoms, thereby making it difficult to predict whether the electroactive sites are accessible, partially accessible or inaccessible. Still, the determined TOF values can be useful and relevant for comparisons between similar materials.

- *Stability*

Long-term stability under both operating and shutdown conditions is also crucial when evaluating the performance of HER electrocatalysts. To quantify the long-term operability of the catalyst, the measurement of current density variation with time or by performing CV or LSV for at least 5000 times can be performed.[7,11] In the first case, a current density larger than 10 mA cm^{-2} should be set with an experiment time higher than 10 h.

7.2.2 Oxygen Evolution Reaction

The OER is the anode reaction in the water splitting process and involves the production of oxygen. The commonly accepted mechanisms for the OER comprise (i) a complex multistep proton-coupled electron transfer and (ii) oxygen–oxygen bond formation; and these involve several surface adsorbed intermediates.

Various research groups have proposed possible mechanisms for the OER for either alkaline or acid conditions, presenting similarities as well as discrepancies.[12] The majority of the proposed mechanisms involve the intermediates MOH and MO, where M represents the metal surface of the catalyst.

However, the main difference between them is the actual reaction that leads to the formation of O_2. Generally, there are two distinct routes to form O_2 from a MO intermediate.[12] The first involves the direct combination of two MO to produce oxygen ($2MO \rightarrow 2M + O_2(g)$), while the second comprises the formation of MOOH as an intermediate ($MO + H_2O(l) \rightarrow MOOH + H^+ + e^-$ for acidic solutions and $MO + OH^- \rightarrow MOOH + e^-$ for alkaline solutions). This MOOH intermediate is then decomposed to $O_2(g)$. (In acidic solutions: $MOOH + H_2O(l) \rightarrow M + O_2(g) + H^+ + e^-$. In alkaline solutions: $MOOH + OH^- \rightarrow M + O_2(g) + H_2O(l)$.) Regardless of the differences, it is commonly accepted that in the OER the bonding interactions within the intermediates are critical for the overall electrocatalytic performance.

The reaction rate will depend on the kinetic constraints of the basic reaction steps, which in turn depend on the relative stability of each intermediate (MOH, MO, and MOOH) and the activation barriers between them. This often results in sluggish kinetics and large overpotential.[13] To date, noble-metal-based materials such as RuO_2 and IrO_2 electrocatalysts show the highest OER activity with IrO_2 showing the best performance/stability relation.

To evaluate the performance of OER electrocatalysts, there are some important parameters that must be determined, such as the aforementioned overpotential, exchange current density, Tafel slope, TOF and long-term stability.[3,7] As for the HER, the overpotential is also set at a current density of $j = 10\ mA\ cm^{-2}$.

7.2.3 Oxygen Reduction Reaction

The oxygen reduction reaction (ORR) is the most important reaction in life-related processes such as biological respiration. It is also critical in electrochemical energy systems such as fuel cells and metal–air batteries. In aqueous solutions, the ORR can theoretically occur predominantly by two pathways as described in Table 7.1: (a) the direct 4-electron reduction pathway with the production of water in acidic electrolytes or hydroxyl group in basic electrolytes, or (b) the 2-electron reduction pathway where O_2 is reduced to H_2O_2 or HO_2^- (in acidic or basic electrolytes, respectively).[7,14] Usually, the 4-electron direct pathway is highly preferred in fuel cells and the 2-electron reduction pathway is mainly used in the H_2O_2 production industry.

Table 7.1 lists the typical ORR processes along with their corresponding thermodynamic electrode potentials at standard conditions. As it can be observed, the mechanism of the ORR is quite complex involving various intermediates, mainly depending on the nature of the electrocatalyst and the electrolyte used during the reaction. In addition, the sluggish kinetics of the ORR makes the use of an electrocatalyst imperative. Platinum-based materials are still the most active electrocatalysts, however, these suffer from scarcity, high cost and low durability, which limits their large-scale application in electrochemical energy systems.[7,15]

Table 7.1 Possible ORR reaction pathways and the respective thermodynamic electrode potentials.[3,7]

Electrolyte	Reaction	Reduction potential (V *vs.* NHE)
Alkaline (aqueous)	**4-Electron pathway:**	
	$O_2 + 2H_2O + 4e^- \rightarrow 4HO^-$	+0.401
	2-Electron pathway:	
	$O_2 + H_2O + 2e^- \rightarrow HO_2^- + OH^-$	−0.065
	$HO_2^- + H_2O + 2e^- \rightarrow 3HO^-$	+0.867
	Or decomposition:	
	$HO_2^- \rightarrow 2HO^- + O_2$	
Acidic (aqueous)	**4-Electron pathway:**	
	$O_2 + 4H^+ + 4e^- \rightarrow 2H_2O$	+ 1.229
	2-Electron pathway:	
	$O_2 + 2H^+ + 2e^- \rightarrow H_2O_2$	+0.67
	$H_2O_2 + 2H^+ + 2e^- \rightarrow 2H_2O$	+1.77
	Or decomposition:	
	$2H_2O_2 \rightarrow 2H_2O + O_2$	

To evaluate newly developed electrocatalysts, certain electrochemical parameters are measured to establish benchmarks:[7,15] the onset potential (E_{onset}), which is defined as the potential at which the reaction starts, the diffusion-limited current density (j_L), which is the current density reached when the reaction at the electrode is totally controlled by mass transport (a plateau is reached and $j = j_L$), the kinetic current density (j_k) defined as the current in the absence of mass-transfer limitations, and the number of electrons transferred per O_2 molecule (nO_2). The nO_2 provides information about the ECs' selectivity for one of the possible reaction pathways (2- or 4-electron) and is generally determined using the Koutecky–Levich (K–L) equation, from the slopes of the j^{-1} *vs.* $\omega^{-1/2}$ plot (Figure 7.3).[16] Other parameters discussed in Section 7.2.1 like TS and long-term stability are also evaluated. Ideally, an optimal electrocatalyst should have a high j_L value, less negative E_{onset}, low TS and selectivity for the 4-electron pathway.

7.3 Transition Metal-based Foams as HER Electrocatalysts

Transition metal-based foams have attracted considerable attention in HER electrocatalysis due to their availability, excellent electrocatalytic performance and high stability. The inherent electrocatalytic activities of nickel, copper and iron elements makes them auspicious foams to be applied as both electrode support materials and as electrocatalysts. In addition, transition metal oxides based on Ni, Cu, Fe, Co, Mo and W, as well as their derivatives, can be loaded or grown on the previous foams leading to enhanced electrocatalytic performances.

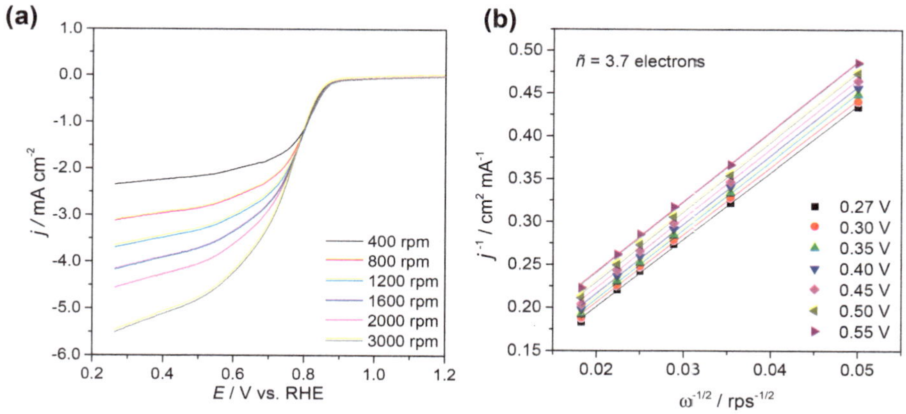

Figure 7.3 ORR LSV curves of a Mn/N_3S_3-GF modified electrode, acquired at different rotation rates in O_2-saturated 0.1 mol dm^{-3} KOH solution at 0.005 V s^{-1} (a) and the respective Koutecky–Levich (K–L) plots obtained from LSV data (b). Reproduced from ref. 16 with permission from Elsevier, Copyright 2019.

7.3.1 Nickel Foam as Support for Metal Oxide Electrocatalysts

In 2015, Jin *et al.*[17] reported the preparation of a nanoflower-like metallic conductive MoO_2 on Ni foam (NFL MoO_2/NF) by a simple hydrothermal method followed by annealing treatments (500 °C in a N_2 and 400 °C in a H_2). The material was then applied as an integrated cathode towards the HER and obtained η_{10} = 55 mV *vs.* RHE, j_0 of ~1.8 mA cm^{-2} and good stability. The authors attributed the superior HER performance to a combination of factors. First, the MoO_2 has good electrical conductivity as a result of its distorted rutile structure with complex bonding involving delocalization of some of the Mo electrons in the conduction band. Second, the high surface area of MoO_2/NF nanoflower-like structure corresponding to many active sites exposed for the HER. Third, the porous configuration of Ni foam allowing electrolyte diffusion and, finally, the NFL structure that avoids the use of polymer binders.

7.3.2 Nickel Foam as Support for Metal Sulfide Electrocatalysts

Considering that metal oxides have lower conductivity compared to metal sulfides, different research groups have reported Ni–S alloys on Ni foams with excellent HER activities.[18,19] Ni–S alloys have been recognized not only as excellent electrocatalysts for the HER but also for the OER because of their facile preparation, good electrical conductivity, excellent catalytic ability and low cost. A good example is Ni_3S_2, a metal chalcogenide that occurs naturally as the mineral heazlewoodite. Due to the continuous network of Ni–Ni bonds throughout its structure, it exhibits intrinsic metallic behaviour. This material has been widely used as electrocatalysts for the HER and OER.[18–22] Sun and co-workers[18] reported the fabrication of a nickel sulfide nanosheet array supported on Ni foam (Ni_3S_2/Ni) by a one-step hydrothermal treatment of commercial porous Ni foam in the presence of $Ni(NO_3)_2·6H_2O$ and thiourea. The developed Ni_3S_2/Ni foam was applied as a HER electrocatalyst in both basic (1.0 M KOH) and neutral (1 M PBS) solutions with η_{10} = 123 mV and η_{100} = 260 mV *vs.* RHE in basic and η_{10} = 220 mV and η_{100} = 396 mV *vs.* RHE in neutral conditions. The Tafel slopes obtained were 110 and 118 mV dec^{-1} in basic and neutral media, respectively. In 2015, Ouyang *et al.*[19] fabricated hierarchically a porous Ni_3S_2 nanorod array foam by a simple hydrothermal method assisted by the pre-treatment of the Ni foam with HCl (Ni_3S_2/AT-Ni foam) using Ni foam and thiourea as Ni and S sources, respectively. The as-prepared Ni_3S_2/AT-Ni foam exhibited excellent HER activity with η_{10} = 200 mV *vs.* RHE and a TS of 107 mV dec^{-1} when compared to Ni_3S_2/Ni foam (without HCl treatment) with η_{10} = 300 mV *vs.* RHE and a TS of 141 mV dec^{-1}. In addition, the Ni_3S_2/AT-Ni foam presented superior stability with no significant variations in η_{10} after continuous CV scanning at a scan rate of 50 mV s^{-1} for 1000 cycles. In this study, several important aspects were pointed out. The pre-treatment by HCl was important to activate the Ni foam surface and to achieve a high density of hierarchically porous Ni_3S_2 nanorod arrays on it, as well as the desired structure and morphology. The authors attributed the excellent performance of this system to the nanorod array structure, which provides fast channels for both electron transfer and mass transport. Additionally, the large electrochemical surface area, resulting from the unique hierarchically porous structure and the increased number of exposed active sites for reaction, plays a crucial role in the enhanced electrocatalytic activity. Finally, the direct growth of Ni_3S_2 nanorod arrays with Ni foam as the Ni source provides a strong combination interaction between the Ni foam and Ni_3S_2 enhancing the long-term stability. Zou and co-workers[21] have also grown Ni_3S_2 nanosheet arrays on nickel foam (Ni_3S_2/NF). The synthetic method used was a direct sulfurization of Ni foam with thiourea as the source of sulfur, in a hydrothermal system at 150 °C for 5 h, which led to the formation of ($\bar{2}\bar{1}0$) high index faceted Ni_3S_2 nanosheet arrays. The as-prepared Ni_3S_2/NF was applied as a HER electrocatalyst in both neutral and alkaline solution, and as an OER catalyst in alkaline solution (see Section 7.3). In neutral electrolyte it presented η of 170 mV *vs.* RHE while in alkaline the value obtained was 223 mv *vs.* RHE (Figure 7.4). Apart from the excellent electrocatalytic performance, it presented an exceptional stability with no significant losses of HER current density after continuous operation at η = 170 mV for 200 hours. This was further confirmed by comparing the polarization curves obtained for cycle 1 and 10 000, which were practically equal. Furthermore, the authors evaluated the morphology and microstructure of Ni_3S_2/NF by SEM, TEM and XPS analysis after these tests and observed that it was intact. The electrocatalyst's excellent performance was attributed to a combination of factors: (1) Ni_3S_2 is intrinsically metallic so the catalytically

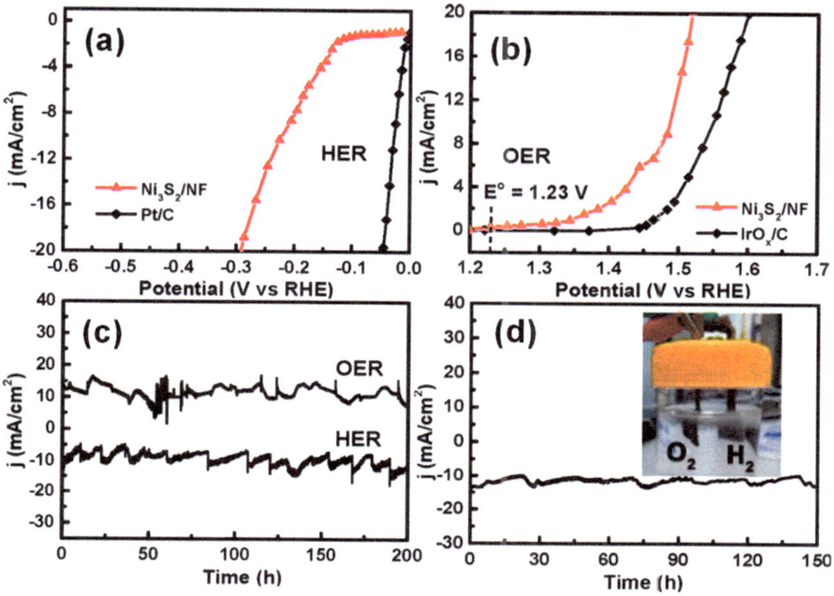

Figure 7.4 Steady-state current density as a function of applied voltage in alkaline media (pH 14) over Ni_3S_2/NF in the (a) HER and (b) OER. (c) Current density *vs.* time (*I–t*) curves of the HER and OER with Ni_3S_2/NF recorded for over 200 h. (d) *I–t* curve at pH 14 at an applied potential of 1.76 V for an electrolyzer whose anode and cathode both constitute Ni_3S_2/NF. Inset: A photograph showing the generation of H_2 and O_2 bubbles on the Ni_3S_2/NF electrodes. Reproduced from ref. 21 with permission from American Chemical Society, Copyright 2015.

active sites are easily accessible to electrons, (2) the nanosheet array of Ni_3S_2/NF contributes to a larger electroactive area for the reaction to occur, (3) the strong interaction between the Ni foam and the Ni_3S_2 contributes not only to an easy electron transfer between both but also to an increased stability, (4) the $(\bar{2}\bar{1}0)$ high-index facets of Ni_3S_2 have a more favourable surface structure for electrocatalysis than the low-index.

Zhu and co-workers[23] developed a NiS microsphere film grown *in situ* on Ni foam (NiS/Ni foam) *via* a sulfurization reaction. The authors used the direct sulfurization of bare Ni foam (Figure 7.5(a)) with sulfur powder which led to an *in situ* growth of a NiS microsphere film on the Ni foam, as depicted in Figure 7.5(b) and (c). This NiS/Ni foam electrode was applied as a bifunctional electrocatalyst for overall water splitting. Regarding its performance as a HER electrocatalyst, η_{20} of 158 mV *vs.* RHE was obtained as well as a Tafel slope of 83 mV dec^{-1}. It also showed superior stability with no significant losses after both CV cycling for 1000 cycles at v = 100 mV s^{-1} and the continuous electrolysis at η of 148 mV for 20 h. More importantly, when applied as a bifunctional electrocatalyst, it enabled a high-efficiency alkaline water

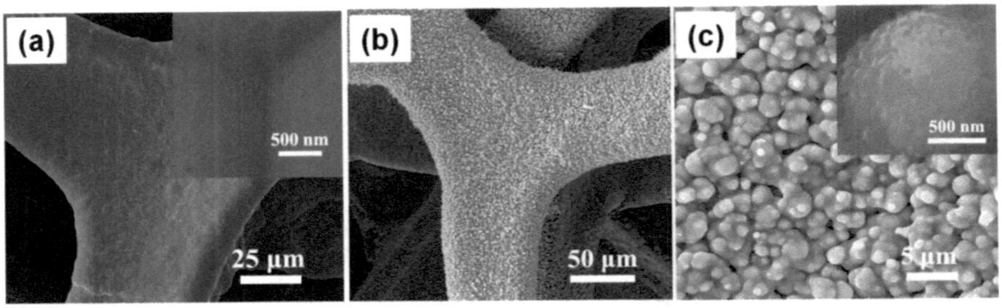

Figure 7.5 SEM images of bare Ni foam (a) and of NiS/Ni foam (b and c). Reproduced from ref. 23 with permission from the Royal Society of Chemistry.

electrolyzer with a cell voltage of 1.64 V (for $j = 10$ mA cm^{-2}), creating a promising alternative for water splitting devices for large-scale production of H_2.

Later, Cao *et al.*[24] reported a new nanostructure with 1D molybdenum sulfide/nickel sulfide (MoS_2/Ni_3S_2) nanoarrays directly grown on a 3D Ni foam by a one-step hydrothermal method using thiourea and sodium molybdate. The prepared material showed excellent HER activity with η_{10} of 76 mV *vs.* RHE. In the same experimental conditions, Ni_3S_2/Ni and Ni foam gave 260 mV and 350 mV *vs.* RHE, respectively. Similarly, the Tafel slope obtained for MoS_2/Ni_3S_2 (56 mV dec^{-1}) is lower than that for Ni_3S_2/Ni (95 mV dec^{-1}) and Ni foam (115 mV dec^{-1}). This excellent performance was attributed to the high number of active sites, the super hydrophilic interface, the electrical coupling between MoS_2, Ni_3S_2 and Ni foam and the synergetic effects generated by the MoS_2/Ni_3S_2 nanoarrays. A cobalt sulfide/nickel sulfide heterostructure has also been hydrothermally grown on a nickel foam (CoS_x/Ni_3S_2@NF) and applied as both a HER and OER electrocatalyst.[25] The electrocatalyst needed 204 mV *vs.* RHE to reach a current density of 10 mA cm^{-2} while CoS_x@NF and Ni_3S_2@NF needed 240 and 269 mV *vs.* RHE, respectively. Tafel slopes were also determined with values of 133, 136 and 150 mV dec^{-1} for CoS_x/Ni_3S_2@NF, CoS_x@NF and Ni_3S_2@NF, respectively. Similar to the previous example, the combination of two different metal sulfides led to an increase in the number of active sites and, consequently, to a better electrocatalytic activity. Furthermore, the formation of a Co–S–Ni moiety improved charge-transfer efficiency. Shen and co-workers[26] have also prepared starfish-like Co–Ni–S nanosheets supported on porous nickel foam (Co–Ni–S/NF) *via* a one-step hydrothermal method using cobalt(II) chloride hexahydrate, thiourea and Ni foam as Co, S and Ni source, respectively. Initially, the shape of the sample was adjusted by controlling the reaction time (7 h, 11 h and 15 h). After 11 h, the sample (Co–Ni–S/NF-11) displayed the most regular structure of nanosheet leading to the best HER electrocatalytic activity with η_{100} of 284.3 mV *vs.* RHE (375.6 and 350.3 mV for Co–Ni–S/NF-7 and Co–Ni–S/NF-15, respectively) and a TS of 139.2 mA dec^{-1}. The authors further evaluated the effect of P doping on Co–Ni–S/NF-11 using sodium hypophosphite as P source (3.5, 7 and 10.5 mmol). The intermediate amount used led to a material (P–Co–Ni–S/NF) with the best HER performance, namely, η_{100} of 187.4 mV *vs.* RHE and a TS of 125.3 mV dec^{-1} while $P_{3.5}$–Co–Ni–S/NF and $P_{10.5}$–Co–Ni–S/NF showed η_{100} of 266.5 mV and 196.6 mV, respectively. In terms of HER stability, the P–Co–Ni–S/NF catalyst had a better performance than Co–Ni–S/NF-11 requiring an overpotential of 190 mV, 90 mV smaller than that of Co–Ni–S/NF-11, to drive a stable j of 10 mA cm^{-2} for over 16 hours.

The hydrothermal method was also used by Sivanantham *et al.*[27] to prepare a novel bifunctional HER and OER electrocatalyst by the *in situ* growth of hierarchical $NiCo_2S_4$ nanowire arrays on a Ni foam substrate. The authors followed a two-step approach (Figure 7.6): First, Ni–Co-carbonate hydroxide nanowire arrays were uniformly grown on a Ni foam by hydrothermal treatment of Co^{2+} and Ni^{2+} with water in the presence of urea at 120 °C for 8 h. Then, the Ni–Co-carbonate hydroxide nanowire arrays were fully converted with spinel $NiCo_2S_4$ nanowire arrays, according to a coordinating etching and precipitating process in a sodium sulfide environment at 160 °C for 8 h (Figure 7.6(b)). The second step (anion exchange reaction) involved the replacement of the CO_3^{2-} and OH^- anions of the Ni–Co-carbonate hydroxide nanowire arrays by the active sulfur anions (S^{2-}) released from sodium sulfide to form hierarchical $NiCo_2S_4$ nanowire arrays on the Ni foam. The exchanged CO_3^{2-} and OH^- anions reacted with the H^+ ions in the solution to

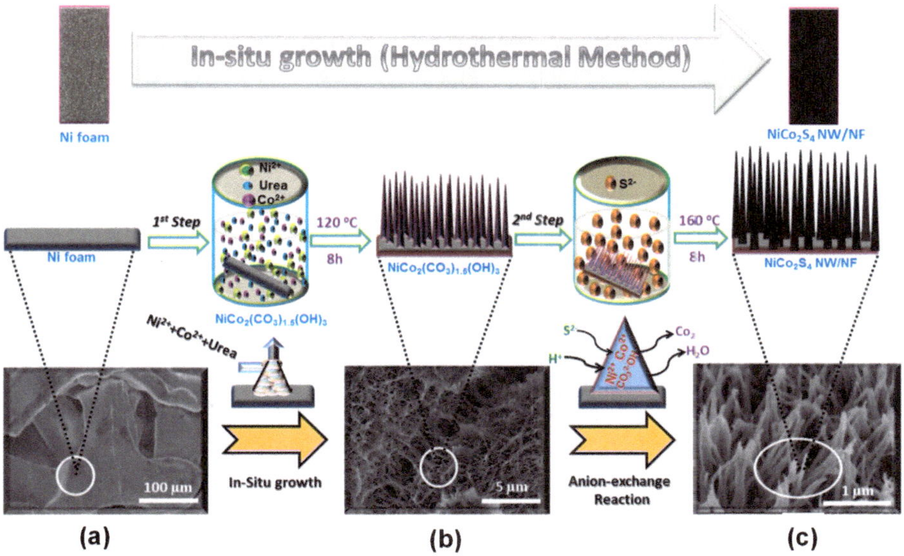

Figure 7.6 Schematic illustration of the formation of $NiCo_2S_4$ nanowire arrays on Ni foam and their corresponding morphology. (a) Bare Ni foam substrate, (b) *in situ* growth of $NiCo_2(CO_3)_{1.5}(OH)_3$ nanowire arrays on Ni foam (1st step), (c) anion exchange reaction with complete growth of hierarchical $NiCo_2S_4$ nanowire arrays on Ni foam (2nd step) using a hydrothermal method. Reproduced from ref. 27 with permission from John Wiley and Sons, Copyright 2016.

form CO_2 and H_2O. The Ni foam acted as a skeleton that anchored the $NiCo_2S_4$ nanowire arrays. The resulting Ni foam surface was fully black (no patches), which suggested the uniform formation of $NiCo_2S_4$ nanowire arrays on the Ni foam (Figure 7.6(c)). The resulting material ($NiCo_2S_4$ NW/NF) exhibited η_{10} of 210 mV *vs.* RHE, which was significantly superior to the values obtained for $NiCo_2S_4$, Ni_3S_2/NF and $NiCo_2O_4$/NF (240, 310 and 310 mV, respectively). This catalyst showed good stability at a static potential of −0.323 V *vs.* RHE for 50 h maintaining 87% of its initial current density.

$NiCo_2S_4$/Ni foam electrodes have also been prepared by Ma *et al.*[28] through the complete sulfidation of networked ultrathin NiCo-layered double hydroxide nanoflakes grown on Ni foam (NiCo-LDH/Ni foam). The preparation involved two steps: (1) hydrothermal synthesis where Ni foam was immersed into an aqueous solution containing Ni^{2+} and Co^{2+}. The formation and uniform assembly of NiCo-LDH nanoflakes on the entire surface of the Ni foam was facilitated by the gradual release of OH^- ions due to the hydrolysis of hexamethylenetetramine (HMT) at high temperature; (2) sulfurization process where the as-prepared NiCo-LDH nanoflakes were converted to porous and ultrathin $NiCo_2S_4$ nanoflakes with well-retained morphology. The as-prepared $NiCo_2S_4$/Ni foam showed remarkable HER activity in alkaline medium presenting η_{10} of 65 mV *vs.* RHE (140 mV for NiCo-LDH/Ni foam and 240 mV for bare Ni foam) and a TS of 84.5 mV dec^{-1} (124.2 and 328.5 mV dec^{-1} for NiCo-LDH/Ni foam and bare Ni foam, respectively). These remarkable results were attributed to the ultrathin and porous $NiCo_2S_4$ nanoflakes' architecture that can provide many exposed active sites, the highly conductive Ni foam that promoted the transfer of electrons, and to the 3D-networked structure that facilitated the diffusion and penetration of electrolyte. Liu *et al.*, reported, in 2017, the preparation of $NiCo_2S_4$@NiFe LDH/Ni foam heterostructures by a three-step hydrothermal approach.[29] The authors combined experimental research with first-principles calculations based on density functional theory (DFT) to engineer the $NiCo_2S_4$@NiFe LDH heterostructures' interface for enhancing overall water splitting performance. The $NiCo_2S_4$@NiFe LDH/Ni foam exhibited η_{10} of 200 mV smaller than those obtained for $NiCo_2S_4$/Ni foam and NiFe LDH/Ni foam (248 mV and 225 mV, respectively) but much higher than that obtained for 20% Pt/C (40 mV). Furthermore, the Tafel slopes obtained were 48, 113.7, 85.8 and 101.1 mV dec^{-1} for 20% Pt/C, $NiCo_2S_4$/Ni foam, NiFe LDH/Ni foam and $NiCo_2S_4$@NiFe LDH/Ni foam, respectively.

7.3.3 Nickel Foam as Support for Metal Selenide Electrocatalysts

Besides sulfurization, direct selenation of foams has also been used to prepare HER electrocatalysts. Sun and co-workers[30] developed a NiSe nanowire film on nickel foam (NiSe/NF) by *in situ* hydrothermal treatment of NF using NaHSe as the Se source. This 3D electrode showed high HER activity requiring an overpotential of 96 mV to achieve 10 mA cm^{-2} and showed a Tafel slope of 120 mV dec^{-1}. The NiSe/NF also presented good stability with η_{20} only increasing by 7

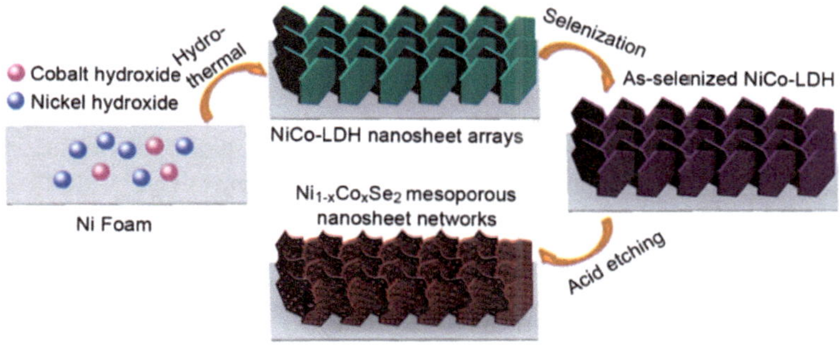

Figure 7.7 Schematic illustration of the synthesis of 3D $Ni_{1-x}Co_xSe_2$ MNSN on Ni foam. Reproduced from ref. 31 with permission from John Wiley and Sons, Copyright 2017.

mV after 1000 cycles between +0.20 and −0.20 V *vs.* RHE. The long-term electrochemical stability was further evaluated by 12 h water bulk electrolysis with the potential required to deliver 50 mA cm^{-2} shifting from −0.182 to −0.188 V.

Liu *et al.*[31] synthesized a series of 3D $Ni_{1-x}Co_xSe_2$ mesoporous nanosheet networks on Ni foam ($Ni_{1-x}Co_xSe_2$ MNSN/NF with $0 \leq x \leq 0.35$) through the topological transformation of NiCo LDH nanosheet arrays by a hydrothermal process and subsequent selenization and acid etching with 0.1 M hydrochloric acid aqueous solution (Figure 7.7). Among the prepared materials, $Ni_{0.89}Co_{0.11}Se_2$ MNSN/NF revealed the best performance with a good electrocatalytic activity and exceptional stability in a wide pH range. In alkaline medium (1.0 M KOH), this material presented an overpotential of η_{10} = 85 mV *vs.* RHE and a Tafel slope of 52 mV dec^{-1}, while in acid medium (0.5 M H_2SO_4) there was an improvement of the performance with η_{10} = 52 mV *vs.* RHE and a Tafel slope of 39 mV dec^{-1}. The electrocatalytic activity was also evaluated in neutral medium (1.0 M PBS), where it presented η_{10} of 82 mV and a Tafel slope of 78 mV dec^{-1}, outperforming all noble metal-free electrocatalysts ever reported until that date. This electrocatalyst also presented excellent stability at overpotentials of 100 and 150 mV with no significant changes in terms of current density over 30 h. DFT simulations verified that $Ni_{0.89}Co_{0.11}Se_2$ remained metallic with a high electrical conductivity. Moreover, substitutional doping of cobalt at an optimal level facilitated the adsorption of both water and H^+ on the electrocatalyst, which were responsible for its superior HER performance in an all-pH environment.

7.3.4 Nickel Foam as Support for Metal Phosphide Electrocatalysts

Foam-based metal phosphide electrocatalysts have also shown great potential toward the HER and OER in alkaline electrolytes, due to their good electrical conductivity and to the suitable *in situ* growth conditions provided by metal foams. Qiao and co-workers reported an innovative strategy to improve both the electrocatalytic performance and structural stability of

Ni_xP_y/Ni foam based ECs by structure design.[32] A 3D porous Ni/Ni_8P_3 electrode was prepared by sequential treatment of commercial Ni-foam with acid activation, followed by phosphorization. In the first step, the Ni foam was initially etched and then exposed to air to form an oxidized layer (NiO/Ni foam). Then, this was converted into Ni_8P_3/Ni foam by phosphorization. The resultant material acted as a self-supported bifunctional electrode for direct water splitting with excellent electrocatalytic activity toward both the HER and OER in alkaline media. Regarding the HER, it exhibited η_{10} = 130 mV *vs.* RHE and a Tafel slope of 58.5 mV dec^{-1}. More importantly, it showed a stable HER performance with negligible current density degradation over 24 h.

Hierarchical MoP/Ni_2P heterostructures on 3D Ni foam (MoP/Ni_2P/NF) have been reported by Du *et al.*[33] These heterostructures were obtained through an easy hydrothermal method to obtain the Mo-based Ni foam precursor, followed by an *in situ* phosphorization procedure. The authors were able to obtain an optimized electrocatalyst with good HER and OER performance in alkaline electrolyte through the manipulation of the ammonium molybdate concentration (for the preparation of the precursor) and the phosphorization temperature. The MoP/Ni_2P/NF showed low overpotentials of 75 mV and 191 mV to achieve j = 10 mA cm^{-2} and 100 mA cm^{-2}, respectively, demonstrating a highly efficient cathode for generating hydrogen from water. A Tafel slope of 100.2 mV dec^{-1} and an excellent stability over 24 h were also obtained. The superior performance was attributed to the 3D porous conductive substrate, distinctive hierarchical heterostructures and collaborative advantages of bimetallic phosphides.

A similar approach was used by Li and co-workers, who synthesized NiCoP nanowire arrays grown on 3D Ni foam (NiCoP NWAs/NF, where NWAs denote cone-shaped nanowire arrays)[34] by a two-step method (hydrothermal process followed by phosphorization, Figure 7.8(a)) using Ni–Co-carbonate hydroxide precursor nanowires as hard templates. These were applied as HER and OER electrocatalysts. Regarding the HER, the NiCoP–NWAs/NF showed η_{10} = 104 mV and η_{100} = 197 mV *vs.* RHE in alkaline media (1.0 M KOH) and a Tafel slope of 54 mV dec^{-1} (Figure 7.8(b and c)). For comparison, the authors also evaluated the HER performance of Co_2P–NWAs/NF, Ni_2P-NSAs/NF (NSAs – nanosheet arrays) and NiCo-CH-NWAs/NF, which showed η_{10} of 131, 152 and 228 mV. The better performance of NiCoP-NWAs/NF was attributed to several aspects like the special cone nanostructure and bifunctional crystal structure, the presence of a bimetal synergetic effect, the inherent nature of transition metal phosphides, and the fast electron transfer between the catalysts. The electrocatalyst also showed excellent HER stability over 28 h.

7.3.5 Nickel Foam as Support for Composite-based Electrocatalysts

The coordination of two or more metal species has also been drawing increasing attention since they can provide active sites with superior activity and improved electronic conductivity, which are of extreme importance

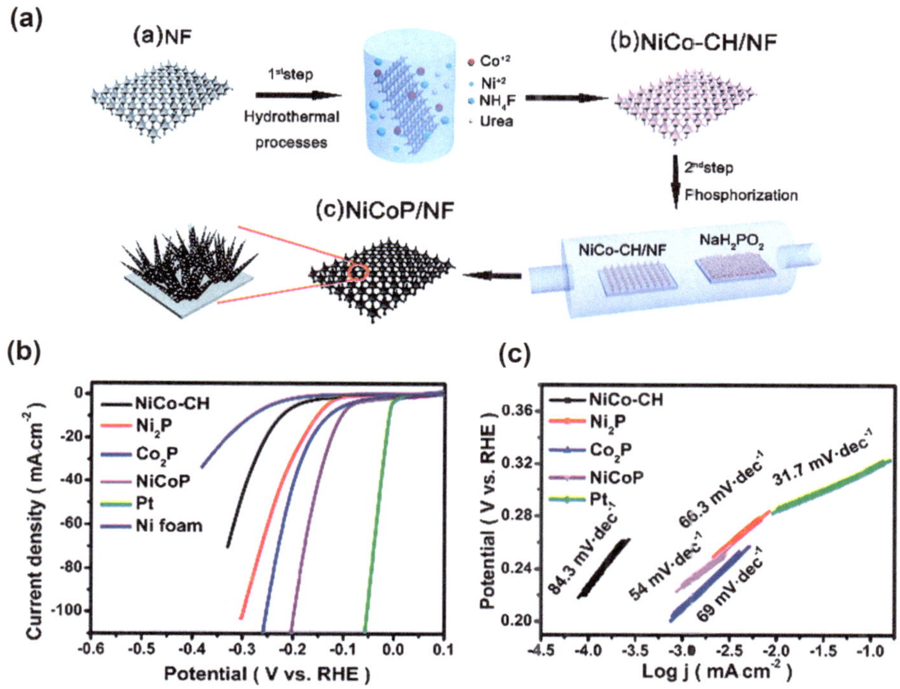

Figure 7.8 Schematic illustration of the two-step method formation of NiCoP cone-shaped nanowire arrays on Ni foam (a), polarization curves of NiCoP-NWAs/NF, Co_2P-NWAs/NF, Ni_2P-NSAs/NF, NiCo-CH-NWAs/NF and Pt/C for the HER at a scan rate of 5 mV s^{-1} (b) and corresponding Tafel plots (c). Reproduced from ref. 34 with permission from the Royal Society of Chemistry.

to electrocatalytic applications. Xiao *et al.*[35] combined 3 transition metals to increase the electrocatalytic performance by taking advantage of the abundant and synergistic interaction between metal–metal and metal–carbon. They synthesized a 3D hierarchical porous catalyst architecture based on Ni, Fe and Co through a hydrothermal deposition and potentiostatic electrodeposition. First, a uniform thin layer of vertically aligned $NiCo_2O_4$ nanoflakes was *in situ* grown onto the nickel foam by a hydrothermal and annealing approach before the topmost NiFe nanosheets were electrodeposited, resulting in the final NiFe/$NiCo_2O_4$/NF electrode. This electrocatalyst showed HER activity with a moderate overpotential of 105 mV to deliver a current density of 10 mA cm^{-2} and a Tafel slope of 88 mV dec^{-1}. In addition, the NiFe/$NiCo_2O_4$/NF electrode showed excellent stability/durability as no obvious degradation after 1000 continuous sweeping cycles between −0.3 and +0.3 V *vs.* RHE at a scan rate of 50 mV s^{-1} or after 10 h at j = 20 mA cm^{-2} was observed. Its good performance was attributed to a combination of different factors, including the unique multilevel hierarchical catalyst architecture.

Ternary NiFeMoS anemone-like nanorods with a scaly surface supported on nickel foam (NiFeMoS/NF-P) have also been synthesized using pluronic P123 (organic) and ammonium molybdate (MoO_4^{2-}, inorganic) as morphology directing reagents.[36] The electrode was prepared by the electrodeposition of Ni–Fe hydroxide nanosheets onto Ni foam followed by a hydrothermal process with MoO_4^{2-}, P123 and thiourea. The organic–inorganic hybrid NiFeMoS/NF-P exhibited a medium electrocatalytic performance, with η_{10} of 100 mV *vs.* RHE and a Tafel slope of 204 mV dec^{-1} in basic electrolyte (1.0 M KOH). This electrocatalyst also showed excellent OER performance (see the next section). In addition, it was used in an alkaline electrolyzer demonstrator for water splitting, achieving a current density of 100 mA cm^{-2} at an applied potential of 1.52 V with excellent durability. Its performance was attributed to a combination of factors like the metallic conductivity of Ni_3S_2 and the bifunctional active sites for the HER and OER. In addition, NiFeMoS/NF-P is directly anchored to the Ni foam without a binder. This direct contact facilitates rapid electron transfer and eliminates interfacial overpotential, resulting in enhanced performance.

Zhang *et al.*[37] reported an iron–nickel nitride nanostructure on surface-redox-etching Ni foam ($FeNi_3N$/NF). This was achieved by, first, growing *in situ* NiFe hydroxide nanosheets ($NiFe(OH)_x$) *via* a simple chemical precipitation reaction and, then, applying thermal ammonolysis, a chemical process based on the reaction with an ammonia-based gas at high temperatures. In this work, Ni foam acted in a dual role as both substrate and slow-releasing nickel precursor. The release of nickel was induced by redox-etching of the iron precursor during the precipitation process. In this process, no nickel precursor or oxidizing agent is needed to produce well-dispersed iron–nickel nitride nanostructures. Another beneficial aspect is that the morphological structure of the $FeNi_3N$/NF increases the specific surface area and mass transfer rate. The $FeNi_3N$/NF electrode exhibited remarkable behaviour for HER in 1.0 M KOH in terms of low overpotentials of 75 mV at 10 mA cm^{-2} and a Tafel slope of 98 mV dec^{-1}, as well as extremely good durability (more than 400 h of constant galvanostatic electrolysis without any visible voltage elevation). This good electrocatalytic performance is achieved by a strong interaction between the electrode patches and active substances, the intrinsic metallic character and the unique structure of $FeNi_3N$/NF.

7.3.6 Other Metallic Foams as Support for HER Electrocatalysis

Besides Ni foams, other metal foams have been used to prepare HER electrocatalysts, even though only a few examples can be found in the literature (see Table 7.2). In 2018, Tang and co-workers[38] reported a simple and efficient method to fabricate a NiCoP@Cu_3P heterostructure on Cu-foam (NiCoP@Cu_3P/CF).$Cu(OH)_2$ nanowires (NWs) were synthesized by direct oxidation directly on Cu-foam. These can be regarded as an extension of Cu-foam, guaranteeing good electrical conductivity. Then, a solvothermal reaction was

Table 7.2 Summary of the HER electrochemical performance (overpotential (η) and Tafel slopes) for nickel, copper and iron foams in alkaline medium (1.0 M KOH).

Active material	Type of foam	η_o (mV)							Tafel (mV dec^{-1})	References
		10 mA cm^{-2}	20	30	50	60	100	200		
NFL MoO_2	Ni foam	55	80						66	17
Ni_3S_2		123					260		110	18
Ni_3S_2/AT		200							107	19
Ni_3S_2		223							—	21
NiS			158						83	23
3D MOS_2/Ni_3S_2		76							56	24
CoS_x/Ni_3S_2		204	280				340	383	133.3	25
Co–Ni–S							284.3		139.2	26
P–Co–Ni–S							187.4		125.3	26
$NiCo_2S_4$ NW		210							58.9	27
$NiCo_2S_4$		65	96						84.5	28
$NiCo_2S_4$@NiFe LDH		200							101.1	29
NiSe		96							120	30
$Ni_{0.89}Co_{0.11}Se_2$ MNSN		85							52	31
NiO		130							59	32
MoP/Ni_2P		75					191		100.2	33
NiCoP-NWAs		104					197		54	34
NiFe/$NiCo_2O_4$		105	140						88	35
NiFeMoS/NF-P		100							204	36
$FeNi_3N$		75							98	37
Co_9S_8@NiCo LDH		168							103	42
FeNiP-NPHC							180		101	43
NiCoP@Cu_3P	Cu foam	54							73	38
Ni–Cr		144							88	39
N7CF-phy	Fe foam				145		200	229	114	40
FeS[a]		63							95	41

[a]Alkaline medium (1.0 M KOH + 0.5 M NaCl).

carried out to obtain NiCo layered double hydroxide (LDH) nanosheets, followed by a phosphating reaction to get NiCoP@Cu_3P/CF. HER studies were conducted in a basic solution (1.0 M KOH) in which NiCoP@Cu_3P/CF exhibited a low overpotential of 54 mV *vs.* RHE at 10 mA cm^{-2} and a small Tafel slope of 72 mV dec^{-1}. The outstanding performance can be ascribed to the large specific surface area, the fast electron transfer and mass transport capability, and the multiple synergetic effects between NiCoP nanosheets and Cu_3P nanowires, and between Ni and Co in the nanosheets. Furthermore, this electrocatalyst also showed excellent OER activity (see Section 7.4).

More recently, in 2020, Qazi *et al.*[39] reported the preparation of a nickel–chromium surface-modified copper foam (Ni–Cr@CF) by an easy annealing process under H_2/Ar atmosphere. Ni–Cr@CF was applied as HER electrocatalysts in basic solution (1.0 M KOH) achieving η_{10} = 144 mV *vs.* RHE and a TS of 88 mV dec^{-1}. The authors also evaluated the stability of the as-prepared Ni–Cr@CF by chronoamperometry at an overpotential of 160 mV for 11 h. No significant changes were observed suggesting good long-term stability. The electrocatalyst also exhibited excellent stability up to 2000 cycles under the basic medium. The authors attributed the efficient HER performance of Ni–Cr@CF to two factors. Firstly, the surface modification of Cu foam with the Ni–Cr alloy, which provided an excellent mechanical adhesion and promoted the electron transfer from copper foam to the active catalytic phase during HER. Secondly, the high electrical conductivity arising from the copper foam which facilitated the fast electron transportation.

A metal iron foam is also an optimal alternative for a highly efficient and economical HER electrocatalyst since it benefits from a three-dimensional skeleton and provides abundant active metal sites; however, it is rarely reported. In 2020, Xiao and co-workers[40] reported an electrocatalyst prepared by a one-step electrochemical erosion method. The iron foam surface was initially oxidized and then, the iron ions were chelated *in situ* with bi-metallic phytates to afford stereo tri-metallic compounds. The prepared N7CF-phy was applied towards the HER in 1.0 M KOH achieving an overpotential of 145 and 229 mV *vs.* RHE for current densities of 50 and 200 mA cm^{-2}, respectively, and Tafel slope of 114 mV dec^{-1}. The use of phytates and bimetal allowed the creation of more electrocatalytic sites and a more complex 3D structure, which enlarged the active surface area and improved the electrocatalytic performance. Very recently, in 2022, Hao and co-workers[41] published a paper on the FeS@IF electrode (where IF stands for Fe foam) for a pH-universal HER. This was achieved by a one-step mild sulfurization etching for 2 h which resulted in an ultrathin FeS nanolayer with fan-like morphology self-supported on a 3D porous iron foam substrate. The as-prepared electrode was able to catalyse the HER in alkaline simulated seawater (1.0 M KOH + 0.5 M NaCl) and neutral electrolyte (1.0 M PBS) with η_{10} of 63 mV and 78 mV *vs.* RHE, respectively. In addition, the FeS@IF electrode continuously catalyses, in 1.0 M PBS, for over 600 h at 0.2–0.4 A cm^{-2} with no significant changes. The authors attributed the excellent performance and robustness to the highly-dispersed nano fan-like structure which enabled a rapid gas release and mass transfer, and to the strong interaction between FeO_xS_y species and the 3D porous substrate.

7.4 Transition Metal-based Foams as OER Electrocatalysts

7.4.1 Metal Oxide-based Electrocatalysts on Nickel Foam

Similar to the case of the HER, transition metal-based foams have attracted considerable attention as electrocatalysts for the oxygen evolution reaction (OER). Many species have been studied as alternatives to the benchmark Ru and Ir-based electrocatalysts. Transition metal oxides and hydroxides of Ni, Co, Mn and Fe have been used due to their high porosity structures and zigzag skeletons, which increase their electrocatalytic surface areas. These morphological features promote an easier penetration of the electrolyte as well as oxygen escape.[44] In 2013, Wang *et al.*[45] reported a unique method to prepare a Ni foam/porous carbon/anodized Ni (NF/PC/AN) electrode for the OER. A homogeneously coated porous carbon membrane derived from ZIF-8 was utilized as an interlayer to protect the inner unstable 3D Ni foam and to support the outermost anodized Ni. Characterization results by Raman and XPS analysis then showed that the anodized Ni was in fact γ-NiOOH. This study presented the first report on a nickel-based OER electrocatalyst, generated *in situ* through the anodization of the innermost Ni foam skeleton with further penetration into the voids and surface coating of the porous carbon membrane. The prepared electrode showed η_{10} of 534 mV *vs.* RHE and high stability. At the same overpotential, the standalone Ni foam only reached approximately 4 mA cm^{-2}. The good OER performance of this novel NF/PC/AN electrode was attributed to its tailored properties, which include the high activity of anodized Ni, an interconnected conductive network and enhanced stability of the electrode.

A novel, simple and scalable leaven dough method was developed by Lu and co-workers to prepare 3D Ni foams. In this, a thin NiO film was grown through thermal oxidation of a Ni substrate to form the final NiO/Ni foam.[46] Briefly, nickel nitrate (Ni precursor and leavening agent) and F127 (solid solvent and fuel) powders were combined by simple grinding and heated at 80 °C. This process led to the formation of a green dough that was calcinated at 300 °C to obtain the porous Ni foam. Then, the NiO/Ni foam was prepared by *in situ* thermal oxidation of the Ni foam. Different temperatures, from 300 to 450 °C, were used and their effect on the OER evaluated. The authors found that an optimal calcination temperature, 350 °C, exists to produce the NiO/Ni composite for OER applications, which showed the lowest η_{onset} and η_{10} values (0.268 and 0.345 V *vs.* RHE, respectively). This outstanding performance of the as-prepared electrocatalyst was attributed to the 3D porous chain-like network of the Ni skeleton and its intact contact with the thin NiO film and to the suitable amount of NiO.

7.4.2 Metal Sulfide-based Electrocatalysts on Nickel Foam

Transition metal chalcogenides have also been used as candidates for OER electrocatalysis with Ni-based compounds being the most promising candidates due to their special e_g orbitals and unique number of 3d electrons.[47] Zhou *et al.* reported,[20] in 2013, a well-defined Ni_3S_2 nanorod array grown on Ni foam (Ni_3S_2/Ni) by a simple hydrothermal method. In the preparation of the electrode, thiourea was used as the S source while Ni foam acted as the Ni source providing the growing sites for Ni_3S_2 nanorods. The electrode showed superior performance with η_{10} of ~187 mV in 0.1 M KOH. In addition to the superior electrocatalytic activity, the Ni_3S_2/Ni electrode also exhibited very good stability; no significant changes were observed in the j–E_p curves even after 1000 potential cycles. Even though the mechanism for the enhanced performance was not totally understood, the authors anticipated that that the rich and unusual valence states of Ni (Ni^0, Ni^+ and Ni^{2+}) in the final electrode had an important role in lowering the overpotential. Also, hydrated nickel oxide derived from the Ni_2S_3 on its surface in basic solution could provide additional active sites.

Table 7.3 summarizes the main OER parameters obtained for different metal foam-based electrodes. Some of the work has already been described in Section 7.3 because the electrodes were used as both HER and OER electrocatalysts. For example, hierarchically porous Ni_3S_2 nanorod array foam (Ni_3S_2/AT-Ni foam), already described in the HER section, has also been applied as an OER electrocatalyst.[19] The Ni_3S_2/AT-Ni foam electrode showed a low overpotential of 217 mV to deliver 10 mA cm^{-2} and a Tafel slope of 163 mV dec^{-1} in 0.1 M KOH. The electrode also exhibited excellent long-term stability with no significant changes in LSVs before and after 1000 cycles at a scan rate of 50 mV s^{-1}.

As mentioned in Section 7.3, Sivanantham *et al.*[27] prepared a bifunctional OER/HER electrocatalyst by the *in situ* growth of hierarchical $NiCo_2S_4$ nanowire arrays on a Ni foam substrate ($NiCo_2S_4$ NW/NF). The prepared electrode exhibited η_{10} of 260 mV *vs.* RHE and a Tafel slope of 40.1 mV dec^{-1}. In the same conditions, the commercial IrO_2 showed an overpotential of 340 mV and a Tafel slope of 72.5 mV dec^{-1}, which is 80 mV higher than that of $NiCo_2S_4$ NW/NF highlighting the electrocatalyst's excellent performance. The enhanced OER activity was attributed to several factors: (1) closely packed arrays of large S^{2-} anions with Ni and Co metal cations in different valence states present in the tetrahedral and octahedral sites, respectively, of the spinel crystal structure, which results in more octahedral active sites of Co(III) cations for an efficient OER at a very low activation potential, (2) good electronic conductivity of Ni which facilitates electron transfer to the substrate, (3) a uniform growth of nanowire arrays on the Ni foam with good separation which resulted in more exposed active sites in electrolyte, (4) the higher electrochemical surface area of $NiCo_2S_4$ NW/NF, (5) the high electrode surface roughness and (6) the absence of binders helped to decrease the charge transfer resistance and increase the electron paths for efficient electron flow between the nanowire array and the substrate.

Table 7.3 Summary of the HER electrochemical performance (overpotential (η) and Tafel slopes) for nickel, copper and iron foams in alkaline medium (1.0 M KOH).

		η (mV)									
	Type of foam	10 mA cm^{-2}	20	30	35	50	60	100	200	Tafel (mV dec^{-1})	References
PC/AN[a]	Ni foam	534								—	45
NiO		345								53	46
Ni_3S_2		187								159.3	20
Ni_3S_2/AT		217								163	19
Mn-Ni_3S_2								347		69.3	52
$NiCo_2S_4$ NW		260								40.1	27
NiSe			270							64	30
NiSe					400					73.3	47
Ni_8P_3										73.2	32
NiFe/$NiCo_2O_4$		240								38.8	35
pc-NiFe-LDH			133					153		30	48
$FeNi_3N$		202								40	37
NiS	Ni foam					335				89	23
CoS_x/Ni_3S_2			280					373	414	105.4	25
Co–Ni–S								296		81.6	26
P–Co–Ni–S								292.2		61.1	26
$NiCo_2S_4$@ NiFe LDH							201			46.3	29
MoP/Ni_2P			309					365		77.6	33
NiCoP-NWAs			270					370		116	34
NiFeMoS/NF-P								270		69	36
Co_9S_8@Nico LDH				278						83	42
NiCoP@Cu_3P	Cu foam	309								45.6	38
Cu oxides/ $Co(OH)_2$ – 8 h						259				58	49
Ni_2P	Fe foam	198									50
N7CF-phy			224					267	275	31	40
Co/Fe^{3+}	Co foam	262						335		62.2	51
Ni/Fe^{3+} foam	Ni–Fe foam	239								70.5	51

[a]Alkaline medium (0.1 M KOH).

7.4.3 Metal Selenide-based Electrocatalysts on Nickel Foam

Similarly to the previous section, direct selenation of foams has also been used to prepare OER electrocatalysts. Sun and co-workers[30] developed a NiSe nanowire film on nickel foam (NiSe/NF) by an *in situ* hydrothermal method. The electrocatalyst exhibited high activity for the OER with an overpotential of 270 mV required to achieve 20 mA cm^{-2} and strong durability in 1.0 M KOH. Characterization results highlighted that the NiOOH species formed at the NiSe surface acted as electrocatalytic sites. The NiSe/NF was also used as a bifunctional electrode and enabled a high-performance alkaline water electrolyzer with 10 mA cm^{-2} at a cell voltage of 1.63 V. Xu *et al.*[47] have also reported the fabrication of hexagonal nickel selenide nanowires on Ni foam (NiSe/NF) using a solution–liquid–solid (SLS) method. The electrode was applied in the OER in 1 M KOH and achieved an overpotential of 0.4 V *vs.* RHE to reach 35 mA cm^{-2}.

7.4.4 Metal Phosphide-based Electrocatalysts on Nickel Foam

Metal phosphides anchored in metal foams have also been used as OER electrocatalysts. As mentioned in Section 7.3, Qiao and co-workers reported the preparation of a 3D porous Ni/Ni_8P_3 electrode by an innovative strategy to prepare a bi-functional electrocatalyst.[32] Regarding the OER, an overpotential of 0.27 V *vs.* RHE was obtained for a current density of 30 mA cm^{-2} and a Tafel slope of 73.2 mV dec^{-1}. The as-prepared electrode also presented excellent stability in 1.0 M KOH at 1.55 V *vs.* RHE with the current density increasing from 41.2 to 48.9 mA cm^{-2} after 24 h. The authors observed that the Ni/Ni_8P_3 electrode showed improved performance in the first 4 h and attributed this enhancement to the increased active electrocatalytic species in 1 M KOH through the conversion of surface Ni/Ni_8P_3 to OER active Ni species (Ni oxides/hydroxides on phosphides). In addition, other factors that can be pinpointed as the reasons for the excellent OER stability are (i) the 3D porous structure which favours oxygen gas release and (ii) the strong contact between catalysts and the Ni-foam as a result of the self-growth character. Other examples include those presented in Table 7.3, whose preparations were already discussed in the previous section.

7.4.5 Metal Composite Electrocatalysts on Nickel Foam

Another efficient way to prepare electrocatalysts with good performances is to combine different metal species. In this sense, in 2016, Xiao *et al.*[35] combined 3 transition metals to achieve a material with increased electrocatalytic OER performance. A 3D hierarchical porous NiFe/$NiCo_2O_4$/NF electrode was prepared by hydrothermal deposition followed by potentiostatic electrodeposition (see Section 7.3 for more details). The electrode showed a remarkable overpotential of 340 mV to deliver an extremely large current density of 1200 mA cm^{-2} and a very low Tafel slope (38.8 mV dec^{-1}), outperforming most state-of-the-art OER catalysts at that time. It also showed excellent stability for 10 h in continuous bulk water electrolysis in 1 M KOH solution at $j = 10$ mA cm^{-2} and 50 mA cm^{-2}. This outstanding performance was attributed to the multilevel hierarchical architecture which enabled both an enhanced accessible surface area and a huge number of active sites for reaction. In addition, the nanoporous structures on each hierarchical level are also well connected to neighbouring structures establishing an integrated electrocatalytic system with efficient electron transfer and overall decreased overpotentials. The fact that no binder is used also contributed to a diminished electrical resistance.

Wang's group[48] reported the fabrication by spontaneous *in situ* growth of partially crystalline NiFe LDH nanosheets on Ni foams at room temperature (pc-NiFe-LDH/NF). Instead of the usually used acidic conditions, the precipitation growth process was conducted in alkaline solution which accelerated the formation of NiFe hydroxide and improved the electrocatalyst's performance. The pc-NiFe-LDH/NF exhibited an excellent OER activity (η_{20} = 133 mV, η_{100} = 153 mV and a TS of 30 mV dec^{-1}), and outstanding long-term stability (with continuous OER operation at 1.609 V for 17 h with an increase of j of 2.6%). NiCo LDH nanosheets, together with Co_9S_8 nanowires and Ni foam, have also been used to fabricate a novel OER electrode by a three-step hydrothermal method.[42] The electrode needed 278 mV to achieve a current density of 30 mA cm^{-2} and exhibited a TS of 83 mV dec^{-1}. More importantly, when used for overall water splitting, it only needs 1.63 V to reach a current density of 10 mA cm^{-2}. The nickel foam had an important role greatly improving the conductivity.

7.4.6 Metal-nitride Electrocatalysts on Nickel Foam

An iron–nickel nitride nanostructure on surface-redox-etching Ni foam ($FeNi_3N$/NF) was reported by Zhang *et al.*[37] and applied as both OER and HER electrocatalysts. The obtained electrode exhibited excellent high activity for the OER with a low overpotential of 202 mV *vs.* RHE at $j = 10$ mA cm^{-2} and a Tafel slope of 40 mV dec^{-1}. The authors also investigated the role of Ni foam on the electrode's fabrication and performance. In this sense, samples decorated with $FeNiN_x$ and Fe_xN on carbon cloth (CC) were also prepared and studied. The pristine CC electrode exhibits an insignificant response toward the OER, as shown in Figure 7.9. Although there is a great improvement of OER activity when CC is modified with Fe_xN, especially with $FeNiN_x$, their performances are still behind $FeNi_3N$/NF, suggesting the importance of using Ni foam on the electrode fabrication.

7.4.7 OER Electrocatalysts on Other Metal Foams

Other metal foams like copper and iron have also been used to prepare OER electrocatalysts, although only a few examples can be found in the literature. As referred to in the previous HER section, in 2018, Tang and co-workers[38]

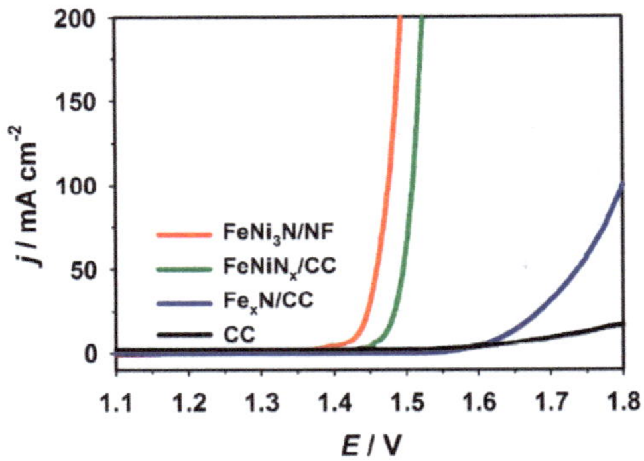

Figure 7.9 Schematic *iR*-corrected polarization curves of $FeNi_3N$/NF, $FeNiN_x$/CC, Fe_xN/CC and CC for the OER in 1.0 M KOH. Reproduced from ref. 37 with permission from American Chemical Society, Copyright 2016.

reported the preparation and application of HER and OER electrocatalysts of a NiCoP@Cu_3P/Cu foam electrode. Regarding the OER, the as-prepared electrocatalyst only needed a small overpotential of 309 mV to reach the current density of 10 mA cm^{-2}. A very small Tafel slope of 45.6 mV dec^{-1} was also obtained indicating that the electrode has remarkable fast charge transfer and reaction kinetics. In 2020, Li's group[49] reported an innovative approach to prepare efficient OER electrocatalysts. They used a femtosecond laser ablation to fabricate rich micro/nanostructures on Cu foam where cobalt hydroxide was then grown *via* the precipitation conversion method. Since morphology can have a huge effect on the electrocatalytic performance, the authors evaluated its evolution throughout the process by SEM. As shown in Figure 7.10(a and b), the initial Cu surface is smooth and, after laser ablation (Figure 7.10(c and d)), it presents a different morphology with rich micro/nanostructures composed of nanoparticles with a size of a few to a few tens of nanometres. Finally, in Figure 7.10(e) and (f), there is a clear presence of layered $Co(OH)_2$ nanosheets after its growth onto the laser fabricated Cu surface. The authors also evaluated different reaction times (2, 4, 6, 8 and 10 h) in order to see its impact on morphology and, in general, how the amount of $Co(OH)_2$ increased by increasing the reaction time. An optimal reaction time of 8 h was achieved. Larger times resulted in $Co(OH)_2$ overgrowth on the surface with clear agglomeration. Among the electrodes evaluated towards the OER, Cu/Cu oxides/$Co(OH)_2$-8 h exhibited the best performance, with an overpotential of 259 mV for j = 50 mA cm^{-2} and a Tafel slope of 58 mV dec^{-1}. Their good performance was attributed to the increased surface area, rich active sites and the superhydrophilicity of the laser produced micro/nanostructures.

Also in 2020, Zhang *et al.*[50] reported the fabrication of Ni_2P nanosheets on iron foam (Ni_2P-FF) by a two-step method to be used as OER electrocatalysts. The resulting electrode showed an overpotential of ~198 mV to reach 10 mA cm^{-2}. The electrocatalyst also showed very good durability

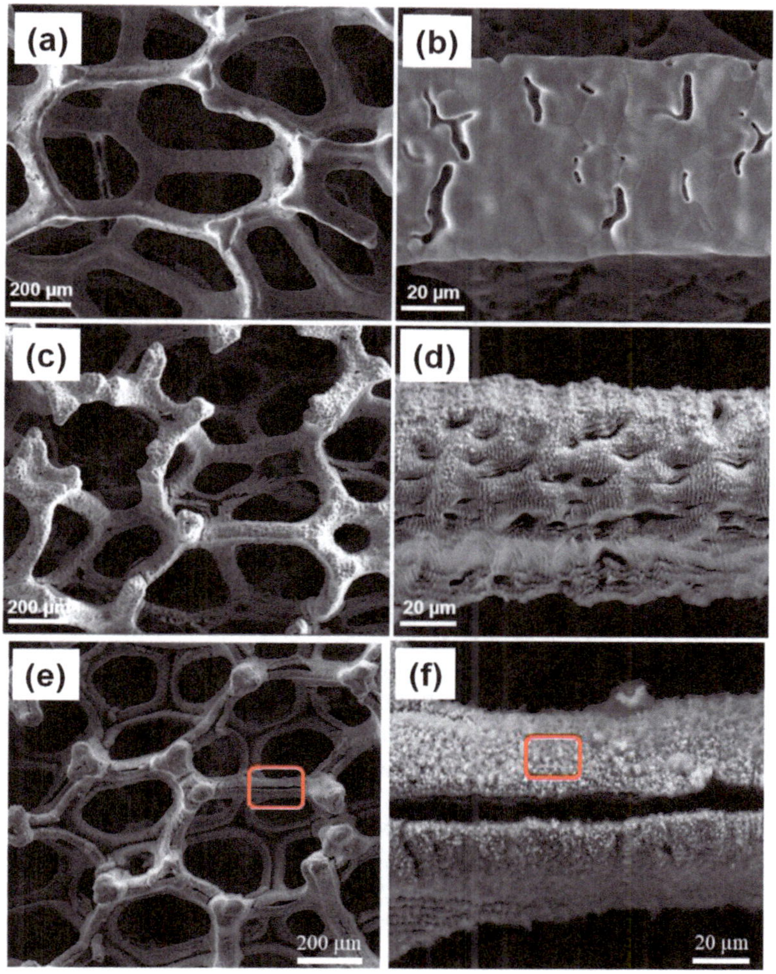

Figure 7.10 SEM images of pristine Cu foam (a and b), after laser ablation (c and d) and after the growth of Cu/Cu oxides/$Co(OH)_2$ for 8 h (e and f). Reproduced from ref. 49, https://doi.org/10.3389/fchem.2019.00900, under the terms of the CC By 4.0 License, https://creativecommons.org/licenses/by/4.0/.

at high current densities. The nanosheet structure of Ni_2P provided more active sites and facilitated the charge transfer during the electrochemical process. In addition, the formation of the phosphate layer optimized the electrode contact and served as a passive layer avoiding iron air oxidation, which resulted in a better electrocatalyst durability. The tight contact and synergistic effect between the two metals (Fe and Ni) was also pointed out as one of the reasons for the electrocatalyst's good performance.

Another group used iron foams to prepare electrodes through *in situ* generation of polymetallic phytates by electrochemical erosion at room temperature.[40] In this work, the iron foam had two purposes, the first to serve as high-surface-area support and the second as a slow releasing iron precursor. The iron was oxidized and, when the Fe^{3+} cations were produced on the metal surface, these were captured by the phytate groups due to the strong chelation capacity towards the metal ions. As a consequence, multi-metallic phytates aggregates were formed with Ni and Co ions. The as-prepared N7CF-phy electrocatalyst with an optimal ratio of Ni : Co of 7 : 1 (where F stands for iron) showed an excellent OER electrocatalytic activity with j_{20} = 224 mA cm^{-2} and a very low Tafel slope of 31 mV dec^{-1}. In addition, it presented a 99% current density retention after 66 and 22 h electrocatalysis at 500 and 100 mA cm^{-2}, respectively.

Zhu *et al.* have also reported a scalable method to produce novel OER foam-based electrodes.[51] The method consisted in the immersion of a commercial metal foam in an aqueous Fe^{3+} solution at room temperature. The metallic foam (Co or Ni) spontaneously reacted with Fe^{3+} and the spontaneously hydrolytic reaction of the metal ions in the presence of air led to the creation of edge-rich bimetal hydroxide species on the metal foam's surface. The prepared electrodes (Co/Fe^{3+} and Ni/Fe^{3+}) showed better performance than the pristine foams. The Co/Fe^{3+} foam electrode required overpotentials of 262 mV *vs.* RHE to achieve a current density of 10 mA cm^{-2} while the Ni/Fe^{3+} foam electrode only needed 239 mV *vs.* RHE. In the same conditions, the pristine cobalt and nickel foams required overpotentials of approximately 344 and 335 mV *vs.* RHE, respectively. Additionally, the Co/Fe^{3+} and Ni/Fe^{3+} foam electrodes showed low Tafel slopes of 62.2 and 70.5 mV dec^{-1}. More importantly, the developed electrodes revealed to be very stable for operation times over 70 h. The excellent performance of Co/Fe^{3+} and Ni/Fe^{3+} foam electrodes was attributed to the unique microstructural feature in the formed hydroxides.

7.5 Transition Metal-based Foams as ORR Electrocatalysts

The design of foam-based electrocatalysts for oxygen reduction is still mainly focused on carbon foams. Usually, they are either doped with different heteroatoms (such as N, S, B and P) or modified through direct growth of transition metal oxides. Only a few examples can be found in the literature concerning the development and application of transition metal-based foams. The first study was reported in 2014 by Modibedi *et al.*[53] The authors synthesized a Pd/Ni foam *via* the electrochemical deposition method, using Cu as a sacrificial metal. The Pd nanostructured catalysts were electrodeposited by surface-limited redox replacement reactions on the Ni foam. This method has the advantage of being conducted at room temperature and requires low concentrations of precursor solutions. It also exhibits the inherent advantages of sequential electrochemical deposition with self-limited deposit growth, namely, fine-tuning of catalytic properties as a function of the ratios between constituent elements by limiting the growth of the deposits at appropriate points. The Pd/Ni foam electrode was evaluated for the ORR in both acidic and basic electrolytes, and, for comparison, the authors also prepared a Pd/carbon paper electrode by a similar procedure. In acid electrolyte (0.1 M $HClO_4$), Pd/Ni foam presented a much higher limiting current density (−2.9 mA cm^{-2}) than Pd/carbon paper (−0.2 mA cm^{-2}). The same behaviour was observed in basic electrolyte (0.1 M KOH) with the Pd/Ni foam showing j_L of −3.03 mA cm^{-2} *vs.* −1.02 mA cm^{-2} obtained for Pd/carbon paper. In addition, in an O_2-saturated atmosphere, the onset potentials of Pd/Ni foam, Ni foam and Pd/carbon paper were −0.0282, −0.272 and −0.0679 V, respectively. These results showed that when Pd is added to Ni foam, there is an improvement of ORR activity. Overall, in alkaline electrolyte, Ni foam proved to be the best choice for depositing Pd for the preparation of an ORR electrocatalyst. The authors attributed these results to the synergistic effect between Pd and Ni.

In 2015, Guo and co-workers[54] reported the application as ORR electrocatalysts of mesoporous Ni-doped Co_3O_4 nanowire (NW) arrays on Ni foam. These were prepared by a hydrothermal method where the Ni foam acted as both the substrate and Ni source. SEM images revealed that the Ni foam was fully covered with the mesoporous Co_3O_4 NWs (Figure 7.11(a)), producing a three-dimensional hierarchical porous structure as (Figure 7.11(a), inset). The Co_3O_4 NWs with sharp tips practically grew vertically to the Ni foam and were well separated (Figure 7.11(b)). The authors also observed that each Ni-doped Co_3O_4 NW contacted with the current collector at the bottom, ensuring the participation of all Co_3O_4 NWs in the ORR, and providing a robust adhesion as well as an efficient route for electron transport. Regarding their application as ORR electrocatalysts (Figure 7.11(c)), these Ni-doped Co_3O_4 NW arrays showed superior performance with a more positive $E_{1/2}$ potential (0.86 V) and a higher diffusion-limiting current density (5.76 mA cm^{-2}) when compared to the Pt/C electrocatalyst (0.85 V, 5.42 mA cm^{-2}).

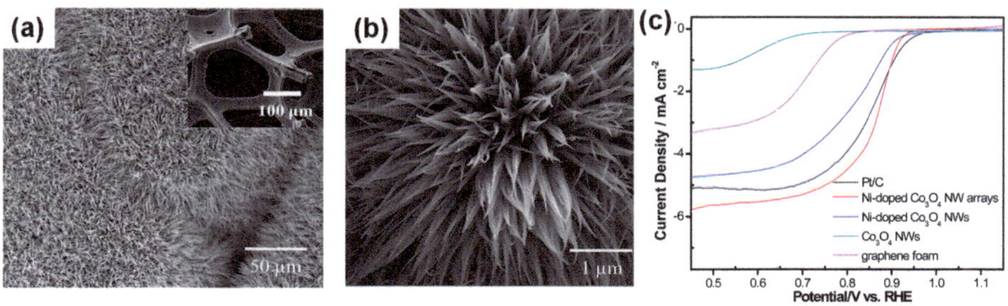

Figure 7.11 Low-magnification SEM images of the Ni-doped Co_3O_4 NW arrays from the top (a and b); the inset shows the 3D hierarchical porous structure, RDE voltammograms at 10 mV s^{-1} for different electrodes in an O_2-saturated 0.1 M KOH solution at a rotation rate of 1600 rpm (c). Reproduced from ref. 54 with permission from the Royal Society of Chemistry.

These also outperformed the Co_3O_4 NWs alone (0.60 V, 1.32 mA cm^{-2}), graphene foam (0.71 V, 3.32 mA cm^{-2}) and Ni-doped Co_3O_4 NWs (0.81 V, 4.82 mA cm^{-2}). Additionally, contrasting with the Pt/C electrocatalyst, the Ni-doped Co_3O_4 NW arrays exhibited high tolerance to methanol and CO. The authors attributed this better performance to the orderly mesoporous structure that provides highly accessible surface areas to the electrolytes and more efficient transport paths for electrons/ions in addition to more active electrocatalytic sites.

Yin *et al.*[52] reported, in 2021, the preparation of a self-supported air electrode with manganese doping nickel sulfide growing on a nickel foam ($Mn\text{-}Ni_3S_2$/NF) by a one-pot hydrothermal reaction. The authors evaluated the effect of the Mn content by assessing three different amounts of $KMnO_4$ (4, 8 and 12 mmol) and the electrocatalysts were labelled as S6M4, S6M8 and S6M12, respectively (S6 stands for 6 mmol thiourea). As it can be observed from Figure 7.12, the $Mn\text{-}Ni_3S_2$/NF electrode with the best ORR performance was S6M8 with the half-wave potential of 0.63 V *vs.* RHE. This electrocatalyst also exhibited the lowest Tafel slope value of 107.5 mV dec^{-1}

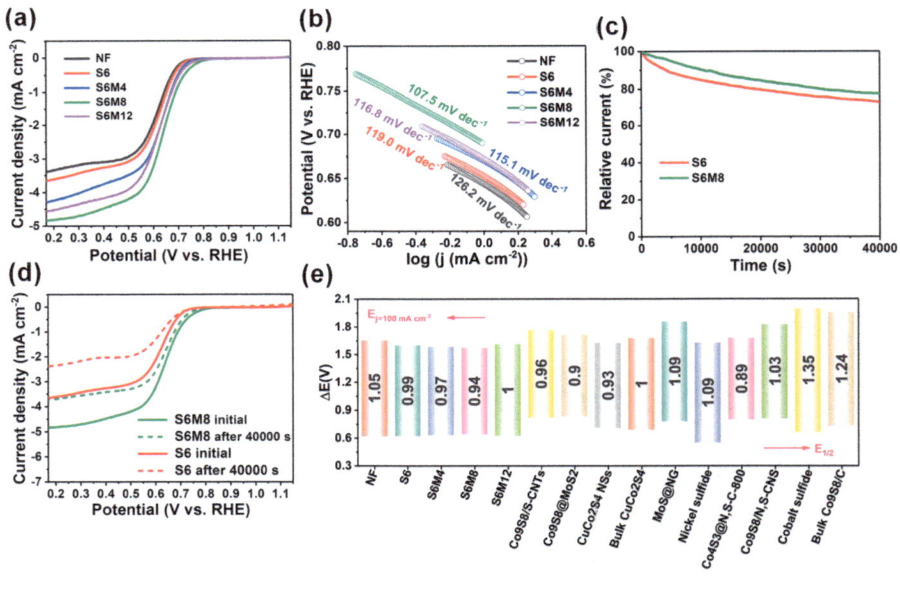

Figure 7.12 LSV curves (a) and Tafel slopes (b) of NF, S6, S6M4, S6M8 and S6M12 toward the ORR in O_2-saturated 0.1 M KOH at a scan rate of 5 mV s^{-1} (rotating speed: 1600 rpm). Chronoamperometric response of S6 and S6M8 at 0.37 V (*vs.* RHE) with a rotation speed of 1600 rpm in 0.1 M KOH (c). LSV curves of S6 and S6M8 toward the ORR before and after the long-term stability test (d). ΔE (the difference between $E_{j=100}$ for the OER and $E_{1/2}$ for the ORR) of the different samples in this work and other works (e). Reproduced from ref. 52 with permission from Elsevier, Copyright 2021.

(Figure 7.12(b)). The stability of S6M8 and S6 (no manganese added) was also assessed by chronoamperometry at 0.37 V *vs.* RHE at 1600 rpm in 0.1 M KOH solution (Figure 7.12(c)). After continuous ORR operation for 40 000 s, the S6M8 retained 77.5% of its initial current while S6 showed lower current retention value (73.1%). The authors also determined the potential gap (ΔE) ($E_{j=100\ \mathrm{mA\ cm^{-2}}} - E_{1/2}$) for the different electrodes (Figure 7.12(e)) since it can be used to assess the bifunctional ability of electrocatalysts for the ORR and OER. Among all the Mn-Ni_3S_2/NF electrodes, S6M8 showed the lowest ΔE value, which suggested that it presented the highest bifunctional ability for oxygen catalytic reactions. In order to clarify why S6M8 showed the best performance amongst the prepared Mn-Ni_3S_2/NF electrodes, the authors used electrochemical impedance spectroscopy (EIS) to estimate their charge transfer resistance (R_{ct}). All the samples showed similar solution resistance (R_s) (about 1.0 Ω); however, regarding R_{ct}, S6M8 showed the smallest value (0.104 Ω) followed by S6M4 (0.118 Ω), S6 (0.135 Ω) and S6M12 (0.158 Ω). These results suggested that there is an optimal Mn doping content (8 mmol) which can lead to a better conductivity and consequently to an improved electrocatalytic performance.

Recently, in 2022, Yu and co-workers[43] reported, for the first time, FeP/$FeNi_2P$ particles embedded in N,P co-doped 1D/3D hierarchical carbon on nickel foam by a hydrothermal-pyrolysis-phosphidation procedure using a metal–organic framework (MIL-FeNi) as precursor. The as-prepared FeNiP-NPHC composite showed superior ORR performance in alkaline seawater with $E_{1/2}$ = 0.83 V and E_{onset} of 0.96 V *vs.* RHE being comparable to the Pt catalyst. The strong coupling effect among the FeP, $FeNi_2P$, and N,P co-doped carbon at the three-phase heterojunction interface was attributed to the regulation of the electronic structure and d-band centre of FeNiP-NPHC and to the 1D/3D hierarchical structure. This effect contributed to expose more active sites, avoid material agglomeration or stacking and speed up the mass transfer process. The FeNiP-NPHC also showed the highest limiting current density of 5.4 mA cm^{-2} at 0.5 V and the lowest Tafel slope (97 mV dec^{-1}), which indicated that the construction of an abundant 3-phase structure contributed to a strong electron interaction and ultimately allowed the acceleration of the reaction rate in the diffusion-controlled and mixed kinetic zone. The proposed electrocatalyst also showed excellent stability with almost no decay in current density after approximately 75 h operation in continuous mode.

7.6 Final Thoughts

Nickel, copper and iron foams with 3D interconnected macroscopic porous structures are ideal growth substrates to load different species like metal oxides, phosphides, chalcogenides and layered double hydroxides since they offer high specific surface areas as well as integrity and electrical conductivity. Engineering of the foams' surface to produce heterostructures led to electrocatalysts with enhanced electrocatalytic performance towards the above-mentioned reaction. Direct sulfurization, phosphorization and selenation of metal foams has also been shown to be an efficient way to produce highly-efficient and robust electrocatalysts for both the hydrogen and oxygen evolution reaction as a result of the formation of films or arrays on the entire surface of the foam. The use of metal foams as both metal source and substrate are two of the strategies established to avoid corrosion and assure electrocatalysis at metal foams in acidic conditions with robustness.

Even though significant progress has been made regarding the development of foam-based electrocatalysts for energy-related reactions there are still some key issues that must be addressed. For instance, the thickness and surface density of commercial metal foams have a great effect on the electrocatalyst loading amount, on the mechanical strength and consequently on the electrocatalytic activity. Another challenge is related to the fact that currently, there is still no accurate methodology established to evaluate the electrocatalytic performance of foam catalysts. The tri-dimensional structure of foams implies that the actual surface area is greater than the geometric area, however, most researchers frequently use the geometric area to calculate electrocatalyst loading which inevitably leads to overvalued performances. This ultimately creates a difficulty in accurately comparing the performance of foam electrocatalysts with those of electrocatalysts loaded onto other substrates. The fact that foams are commonly used as both growth substrate and metal source also brings about some problems as the foam might become fragile due to too high a proportion of foam metal taking part in the synthesis and thus become difficult to handle. Lastly, in the hydrothermal synthesis of transition-metal hydroxide or oxide precursors on metal foams, foam metals like Ni may dissolve out to co-deposit with extrinsic metals to form doping or hybrid structures, but researchers usually don't consider this or they ignore it, which leads to a misunderstanding of the true electrocatalytic active phase.

References

1. N. Armaroli and V. Balzani, *ChemSusChem*, 2011, **4**, 21–36.
2. I. Katsounaros, S. Cherevko, A. R. Zeradjanin and K. J. J. Mayrhofer, *Angew. Chem., Int. Ed.*, 2014, **53**, 102–121.
3. Y. Jiao, Y. Zheng, M. T. Jaroniec and S. Z. Qiao, *Chem. Soc. Rev.*, 2015, **44**, 2060–2086.
4. Z. W. Seh, J. Kibsgaard, C. F. Dickens, I. B. Chorkendorff, J. K. Norskov and T. F. Jaramillo, *Science*, 2017, **355**, eaad4998.
5. X. X. Zou and Y. Zhang, *Chem. Soc. Rev.*, 2015, **44**, 5148–5180.
6. L. Zhang, Q. Chang, H. Chen and M. Shao, *Nano Energy*, 2016, **29**, 198–219.
7. C. Freire, D. M. Fernandes, M. Nunes and V. K. Abdelkader, *ChemCatChem*, 2018, **10**, 1703–1730.
8. M. Zeng and Y. G. Li, *J. Mater. Chem. A*, 2015, **3**, 14942–14962.

9. T. Shinagawa, A. T. Garcia-Esparza and K. Takanabe, *Sci. Rep.*, 2015, **5**, 13801.
10. H. Tao, Y. Gao, N. Talreja, F. Guo, J. Texter, C. Yan and Z. Sun, *J. Mater. Chem. A*, 2017, **5**, 7257–7284.
11. M. S. Faber and S. Jin, *Energy Environ. Sci.*, 2014, **7**, 3519–3542.
12. N. T. Suen, S. F. Hung, Q. Quan, N. Zhang, Y. J. Xu and H. M. Chen, *Chem. Soc. Rev.*, 2017, **46**, 337–365.
13. M. Tahir, L. Pan, F. Idrees, X. W. Zhang, L. Wang, J. J. Zou and Z. L. Wang, *Nano Energy*, 2017, **37**, 136–157.
14. M. H. Shao, Q. W. Chang, J. P. Dodelet and R. Chenitz, *Chem. Rev.*, 2016, **116**, 3594–3657.
15. N. Daems, X. Sheng, I. F. J. Vankelecom and P. P. Pescarmona, *J. Mater. Chem. A*, 2014, **2**, 4085–4110.
16. D. M. Fernandes, P. Mathumba, A. J. S. Fernandes, E. I. Iwuoha and C. Freire, *Electrochim. Acta*, 2019, **319**, 72–81.
17. Y. S. Jin and P. K. Shen, *J. Mater. Chem. A*, 2015, **3**, 20080–20085.
18. C. Tang, Z. H. Pu, Q. Liu, A. M. Asiri, Y. L. Luo and X. P. Sun, *Int. J. Hydrogen Energy*, 2015, **40**, 4727–4732.
19. C. Ouyang, X. Wang, C. Wang, X. Zhang, J. Wu, Z. Ma, S. Dou and S. Wang, *Electrochim. Acta*, 2015, **174**, 297–301.
20. W. Zhou, X.-J. Wu, X. Cao, X. Huang, C. Tan, J. Tian, H. Liu, J. Wang and H. Zhang, *Energy Environ. Sci.*, 2013, **6**, 2921–2924.
21. L. L. Feng, G. T. Yu, Y. Y. Wu, G. D. Li, H. Li, Y. H. Sun, T. Asefa, W. Chen and X. X. Zou, *J. Am. Chem. Soc.*, 2015, **137**, 14023–14026.
22. X. Shang, W. H. Hu, G. Q. Han, Z. Z. Liu, B. Dong, Y. R. Liu, X. Li, Y. M. Chai and C. G. Liu, *Int. J. Hydrogen Energy*, 2016, **41**, 13032–13038.
23. W. Zhu, X. Yue, W. Zhang, S. Yu, Y. Zhang, J. Wang and J. Wang, *Chem. Commun.*, 2016, **52**, 1486–1489.
24. J. Cao, J. Zhou, Y. Zhang, Y. Wang and X. Liu, *ACS Appl. Mater. Interfaces*, 2018, **10**, 1752–1760.
25. S. Shit, S. Chhetri, W. Jang, N. C. Murmu, H. Koo, P. Samanta and T. Kuila, *ACS Appl. Mater. Interfaces*, 2018, **10**, 27712–27722.
26. F. Zhang, Y. Ge, H. Chu, P. Dong, R. Baines, Y. Pei, M. Ye and J. Shen, *ACS Appl. Mater. Interfaces*, 2018, **10**, 7087–7095.
27. A. Sivanantham, P. Ganesan and S. Shanmugam, *Adv. Funct. Mater.*, 2016, **26**, 4661–4672.
28. L. Ma, Yi Hu, R. Chen, G. Zhu, T. Chen, H. Lv, Y. Wang, J. Liang, H. Liu, C. Yan, H. Zhu, Z. Tie, Z. Jin and J. Liu, Self-assembled ultrathin $NiCo_2S_4$ nanoflakes grown on Ni foam as high-performance flexible electrodes for hydrogen evolution reaction in alkaline solution, *Nano Energy*, **24**, 2016, 139–147.
29. J. Liu, J. S. Wang, B. Zhang, Y. J. Ruan, L. Lv, X. Ji, K. Xu, L. Miao and J. J. Jiang, *ACS Appl. Mater. Interfaces*, 2017, **9**, 15364–15372.
30. C. Tang, N. Cheng, Z. Pu, W. Xing and X. Sun, *Angew. Chem., Int. Ed.*, 2015, **54**, 9351–9355.

31. B. Liu, Y. F. Zhao, H. Q. Peng, Z. Y. Zhang, C. K. Sit, M. F. Yuen, T. R. Zhang, C. S. Lee and W. J. Zhang, *Adv. Mater.*, 2017, **29**, 1606521.
32. G. F. Chen, T. Y. Ma, Z. Q. Liu, N. Li, Y. Z. Su, K. Davey and S. Z. Qiao, *Adv. Funct. Mater.*, 2016, **26**, 3314–3323.
33. C. Du, M. Shang, J. Mao and W. Song, *J. Mater. Chem. A*, 2017, **5**, 15940–15949.
34. J. Li, G. Wei, Y. Zhu, Y. Xi, X. Pan, Y. Ji, I. V. Zatovsky and W. Han, *J. Mater. Chem. A*, 2017, **5**, 14828–14837.
35. C. Xiao, Y. Li, X. Lu and C. Zhao, *Adv. Funct. Mater.*, 2016, **26**, 3515–3523.
36. K.-L. Yan, J.-F. Qin, Z.-Z. Liu, B. Dong, J.-Q. Chi, W.-K. Gao, J.-H. Lin, Y.-M. Chai and C.-G. Liu, *Chem. Eng. J.*, 2018, **334**, 922–931.
37. B. Zhang, C. Xiao, S. Xie, J. Liang, X. Chen and Y. Tang, *Chem. Mater.*, 2016, **28**, 6934–6941.
38. X. Ma, Y. Chang, Z. Zhang and J. Tang, *J. Mater. Chem. A*, 2018, **6**, 2100–2106.
39. U. Y. Qazi, R. Javaid, N. Tahir, A. Jamil and A. Afzal, *Int. J. Hydrogen Energy*, 2020, **45**, 33396–33406.
40. X. Chen, Y. Meng, T. Gao, J. Zhang, X. Li, H. Yuan and D. Xiao, *Sustainable Energy Fuels*, 2020, **4**, 331–336.
41. H. Y. Lv, C. Y. Fu, J. L. Fan, Y. R. Zhang and W. J. Hao, *J. Colloid Interface Sci.*, 2022, **626**, 384–394.
42. J. Yan, L. Chen and X. Liang, *Sci. Bull.*, 2019, **64**, 158–165.
43. Q. P. Yu, X. B. Liu, G. S. Liu, X. P. Wang, Z. J. Li, B. Li, Z. X. Wu and L. Wang, *Adv. Funct. Mater.*, 2022, **32**, 2205767.
44. Z. Y. Lu, W. W. Xu, W. Zhu, Q. Yang, X. D. Lei, J. F. Liu, Y. P. Li, X. M. Sun and X. Duan, *Chem. Commun.*, 2014, **50**, 6479–6482.
45. J. Wang, H. X. Zhong, Y. L. Qin and X. B. Zhang, *Angew. Chem., Int. Ed.*, 2013, **52**, 5248–5253.
46. J. Liang, Y. Z. Wang, C. C. Wang and S. Y. Lu, *J. Mater. Chem. A*, 2016, **4**, 9797–9806.
47. K. Xu, H. Ding, K. C. Jia, X. L. Lu, P. Z. Chen, T. P. Zhou, H. Cheng, S. Liu, C. Z. Wu and Y. Xie, *Angew. Chem., Int. Ed.*, 2016, **55**, 1710–1713.
48. H. Li, X. Wang, T. Wang and F. Xiao, *J. Alloys Compd.*, 2020, **844**, 156224.
49. X. Zhou, W. Qi, K. Yin, N. Zhang, S. Gong, Z. Li and Y. Li, *Front. Chem.*, 2019, **7**, 900.
50. M. Zhang, T. Wang, H. Cao, S. Cui and P. Du, *J. Energy Chem.*, 2020, **42**, 71–76.
51. G. X. Zhu, X. Y. Li, Y. J. Liu, Y. Mao, Z. W. Liang, Z. Y. Ji, X. P. Shen, J. Sun, X. F. Cheng and J. J. Mao, *J. Mater. Chem. A*, 2020, **8**, 12603–12612.
52. M. M. Yin, H. Miao, B. Chen, R. G. Hu, L. Xia, C. F. Zhang, F. Wang, H. C. Zhang and J. L. Yuan, *J. Alloys Compd.*, 2021, **878**, 160434.
53. R. M. Modibedi, M. K. Mathe, R. G. Motsoeneng, L. E. Khotseng, K. I. Ozoemena and E. K. Louw, *Electrochim. Acta*, 2014, **128**, 406–411.
54. X. L. Tong, X. H. Xia, C. X. Guo, Y. Q. Zhang, J. P. Tu, H. J. Fan and X. Y. Guo, *J. Mater. Chem. A*, 2015, **3**, 18372–18379.

CHAPTER 8

Challenges and Outlook

Metal foams are an extraordinary class of materials that display several advantages, which can be harnessed to develop high-performance devices. The production routes for these materials essentially require a high throughput at a low energy expense and short preparation times while allowing rigorous control of their key properties. Electrodeposition assumes a prime role among the most reliable and scalable methods to prepare metal foams. However, the range of materials that can be produced in a simple, one-step electrodeposition is rather limited. Nevertheless, this scalable process allows the fabrication of multicomponent metallic foams with different compositions (including bi- and tri-metallic foams). In addition, the pore size, pore distribution and different morphological features can be finely tuned by adapting the synthesis conditions. However, for the time being, liquid metallurgy routes are the prime method for industrial applications due to their manufacturing simplicity and low cost.

The mechanical stability of metal foams is still a key issue that requires careful optimization of the 3D structure *vs.* pore depth, especially when these materials are used as electrodes in catalysers, and in batteries or pseudocapacitors that aim at high volumetric capacity. The need for thinner electrodes that can accommodate sufficient electroactive materials within a restricted volume, while also enduring high mechanical stress, presents a lucrative opportunity for metal foams. Also, their porous structure helps to accommodate an identical mass of active material in thinner layers, enabling reproducible conditions during mass production. The open porous structure accommodates higher mass load and allows thinner layers, increasing the volumetric capacity of the device.

Transition Metal-based Nanofoams for Electrochemical Systems: Manufacturing, Characterization and Applications
By Maria de Fátima Montemor, Diana M. Fernandes, Alberto Adan-Mas, Ana Catarina Alves, Gabriel Garcia Carvalho and Inês S. Marques

Published by the Royal Society of Chemistry, www.rsc.org

Regarding the proper characterization of metal foams, it certainly requires an optimal combination of techniques. While conventional physicochemical characterization techniques provide key information on the chemical composition, morphology and structural features, a major challenge still resides on the accurate estimation of their surface area. Despite several techniques that have been tailored to determine their surface area (*e.g.*, BET), the intricate and complex pore structure still raises some concerns on the accuracy of the determined values. Electrochemical techniques that sense interfacial activity can also bring an important added value to the estimation of the surface area, especially in electrochemical applications. This issue has been explored by many researchers; however, more advances are required to better estimate surface areas from conventional d.c. or a.c. measurements, namely electrochemical impedance spectroscopy.

Undoubtedly, metal foams are key materials to enhance the electrochemical metrics of both pseudocapacitors and batteries. This is clearly confirmed by the industrial commercialization of foams, in particular Ni foams, for battery and supercapacitor applications. Despite the attractiveness of these materials, their optimal loading with active material, their high-power and long-term application, and their functionalization with particular active species still represent a major challenge to introduce real breakthroughs into electrochemical energy storage devices. In supercapacitors, the long-term stability of metal foam-based electrodes is already well established, but the presence of parasitic reactions or additional resistive interfaces still requires important work to further explore their potential as electrode materials for supercapacitor-like applications. In the quest for higher energy density and longer lifetime batteries, metal foams are also a strong enabler of high volumetric energy density. The fabrication of either anodes or cathodes based on metal foams carries the potential for the design of more advanced electrode architectures. These architectures are expected to better accommodate intercalation/deintercalation of Li ions as well as the stress generated by these processes, leading to electrodes with higher rate capacity and reversible capacity on a longer term.

As mentioned in the previous chapter, significant progress has been made regarding the development of foam-based electrocatalysts for energy-related reactions like water splitting and the oxygen reduction reaction. However, there are still some challenges ahead, especially regarding the determination of the true electrocatalytic active phase and the establishment of accurate methodologies to evaluate the electrocatalytic performances of foam electrocatalysts. Aiming at more robust and efficient foam electrocatalysts, researchers should focus on aspects like cleverly using metal foams as metal sources to build foam electrocatalysts with heterostructures and developing transition metal-doped nanoarrays on tridimensional foams for boosted electrocatalytic performances.

Thus, metal foams can pave the way to serve as enablers of different electrochemical energy storage technologies, especially if foams are produced from non-scarce raw materials, and fast, simple and easy to scale-up routes can be implemented. Certainly, metal foams present an exciting and promising field of work. As reviewed in this book, the manufacturing, characterization and applications of transition metal-based nanofoams represent a promising avenue for the development of high-performance electrochemical systems, with the potential to revolutionize the fields of energy storage and catalysis.

Acknowledgements

The authors acknowledge funding from Fundação para a Ciência e a Tecnologia *via* the projects: PTDC/QUI-ELT/28299/2017 and PTDC/QUI-ELT/2075/2020.

This work received support and help from FCT/MCTES (LA/P/0008/2020, DOI: 10.54499/LA/P/0008/2020, UIDP/50006/2020, DOI: 10.54499/UIDP/50006/2020 and UIDB/50006/2020, DOI: 10.54499/UIDB/50006/2020), through national funds.

D. M. F. thanks FCT for funding through the Individual Call to Scientific Employment Stimulus (Ref. 2021.00771.CEECIND/CP1662/CT0007). I. S. Marques thanks FCT for her PhD scholarship 2022.10224.BD. Ana C. Alves acknowledges the PhD fellowship 2020.05407.BD funded by FCT. Gabriel Carvalho thanks the PhD fellowship 2022.13366.BDANA funded by FCT.

Subject Index